UNITEXT for Physics

For further volumes:
http://www.springer.com/series/13351

Carlo M. Becchi · Giovanni Ridolfi

An Introduction to Relativistic Processes and the Standard Model of Electroweak Interactions

Second Edition

Springer

Carlo M. Becchi
Giovanni Ridolfi
Università di Genova
INFN, Sezione di Genova
Genoa
Italy

ISSN 2198-7882 ISSN 2198-7890 (electronic)
ISBN 978-3-319-37441-3 ISBN 978-3-319-06130-6 (eBook)
DOI 10.1007/978-3-319-06130-6
Springer Cham Heidelberg New York Dordrecht London

Softcover reprint of the hardcover 2nd edition 2014

Printed on acid-free paper

Springer is part of Springer Science+Business Media (www.springer.com)

Preface

The natural framework of high-energy physics is relativistic quantum field theory. This is a complex subject, and it is difficult to illustrate it in all its aspects within a normal undergraduate course in particle physics, while devoting a sufficient attention to phenomenological aspects. However, in the small-wavelength limit, the semi-classical approximation is, in many cases of practical relevance, accurate enough to provide reliable predictions without entering the technicalities connected with radiative corrections. In particular, in the framework of the semi-classical approximation it is possible to obtain, in a limited number of pages, the expressions for relativistic cross sections and decay rates in a self-contained and rigorous presentation, starting from the basic principles of Quantum Mechanics. Furthermore, even in the case of the standard model of Electroweak Interactions, the construction of the theory in the semi-classical approximation is exhausted by the study of the classical Lagrangian; many difficult problems, such as those related to the unphysical content of gauge theories, can be dealt with by means of simple prescriptions.

These are the reasons that have determined our choice to base these lecture notes on the semi-classical approximation to relativistic quantum field theory. We believe that this approach leads to a description of the most relevant physical processes in high-energy physics, which is adequate to an undergraduate level course on fundamental interactions.

Of course, the lack of control on radiative corrections has some drawbacks, because they induce transitions which are absent at the semi-classical level. Important examples are the anomalies, the role of intermediate unstable particles and relevant processes such as the Higgs decay into a pair of photons. We have decided to present a simple discussion on each of these subjects, referring the reader to the literature for a complete presentation.

During the preparation of our manuscript we have benefited of the invaluable help and encouragement of Raymond Stora. We are also grateful to Riccardo Barbieri for discussions and suggestions during many years and to our editor Marina Forlizzi for her continuous assistance and friendly advices.

Genoa, February 2014

Carlo M. Becchi
Giovanni Ridolfi

Contents

Chapter 1
Introduction

The study of relativistic processes is based on collision phenomena at energies much larger than the rest energies of the particles involved. In this regime, a large number of new particles is typically produced, with large momenta, or, equivalently, small wavelengths. For this reason, the scheme of ordinary Quantum Mechanics, based on the Schrödinger equation for wave functions that depend on a fixed number of variables, is no longer applicable. A suitable framework is rather provided by electromagnetism, that describes radiation phenomena, and therefore the production and absorption of photons. This analogy leads in a natural way to field theory, in which the dynamical variables that describe a given physical system are fields, i.e. variables labelled by the space coordinates, and independent of each other.

This book aims at presenting a self-contained introduction to the theory of electroweak interactions based on the semi-classical approach to relativistic quantum field theory. This allows us to present a rather detailed discussion of the most relevant aspects of this field.

The main drawbacks of a strict use of this approximation essentially amount to the loss of unitarity, that is of the conservation of probability, in the interaction processes; to the lack of any reference to the problem of anomalies, and the corresponding constraints; and to the lack of any access to processes that appear at the level of radiative corrections, such as the Higgs decay into two photons, and the whole sector of flavour mixing through neutral currents. A systematic analysis of Quantum Field Theory beyond the semi-classical approximation is technically very demanding. An indication of the extension of the theory, at least in the perturbative case, is given at the end of Chap. 4, and important applications, including the possible presence of anomalies, are discussed in Chap. 12. For a more complete discussion we refer to the existing literature; see for example [1–3]. For the same reason, the applications

C. M. Becchi and G. Ridolfi, *An Introduction to Relativistic Processes and the Standard Model of Electroweak Interactions*, UNITEXT for Physics,
DOI: 10.1007/978-3-319-06130-6_1, © Springer International Publishing Switzerland 2014

of Quantum Chromodynamics to problems of phenomenological interest are only marginally discussed; a complete review can be found for example in [4]. However, we consider a self-contained analysis of these subjects very difficult to present in the framework of an undergraduate course.

The text is organized as follows. Chapter 2 is devoted to a review of basic facts in field theory. We illustrate the simple case of spinless neutral particles, described by a single scalar field. We build an action functional for such a system, and derive the corresponding dynamical equations. These are studied explicitly in the small-field regime, that corresponds to asymptotic states in collision processes. Next, we illustrate the role of symmetries in field theories, and their relation with conservation laws. Finally, we illustrate the interpretation of the physical states of a quantum scalar field theory as particle states, and we introduce the concept of antiparticle.

In the following chapter we discuss the calculation of transition amplitudes for collision and decay processes in the semi-classical approximation; we determine the asymptotic conditions to be imposed on the dynamical variables in the remote past and in the far future, and we show how to compute the transition amplitude in the semi-classical limit. We obtain a general formula, which is applied to the case of collisions of spinless particles; the differential cross section for this process is computed in full detail. Chapter 4 is devoted to an introduction to the method of Feynman diagrams. The last section is devoted to the calculation, using the technique of dispersion relations, of a simple diagram with one loop, which is shown to provide the first unitarity correction to the semi-classical approximation to an elastic scattering process.

The results obtained for scalar particles are then extended to spinor particles in Chap. 5, starting from Weyl's basic construction of spinor representations of the Lorentz group. We discuss in some detail the construction of a general class of free Lagrangian densities for spinor fields, including the choices of mass terms that are relevant to neutrino physics.

Spinor fields find their natural application in Quantum Electrodynamics, which is studied in Chap. 6. In this context, we introduce the concept of gauge invariance, which is then extended to the case of non-commutative charges. This allows us to present the Lagrangian density of QCD.

The standard model of electroweak interactions is introduced in Chap. 7. The following chapter is devoted to the description of the Higgs mechanism,which is first illustrated in the simple case of an abelian symmetry, and then extended to the standard model. We have not attempted an exhaustive review of present literature on the experimental foundations of the standard model and of the present status of its experimental tests. The foundations of the standard model are described for example in [5], while for a review of precision tests we refer the reader to [6, 7]. In Chap. 9 we discuss fermion masses and flavour-mixing phenomena. An exhaustive analysis of these subjects can be found in [8, 9]. Finally, in Chap. 10 we present the full Lagrangian density of the standard model, in a form which allows a direct derivation of the Feynman rules.

The full computation of a few decay rates and cross section is presented in Chap. 11. The results obtained are useful, both for the purpose of illustration of the standard computation techniques in field theory, and for their phenomenological relevance.

In Chap. 12 we present two sample calculations beyond the semi-classical approximation. The first example is the decay of a Higgs boson into a pair of photons. The calculation was originally perfomed in ref. [10] in full details; we present here a simplified approach, valid in the small Higgs mass limit, pioneered in ref. [11]. The second example concerns the line shape of a process mediated by the intermediate boson Z^0. An analogous treatment applies to processes mediated by the Higgs and W bosons. In the same chapter we address the problem of anomalous unphysical transitions induced by radiative corrections, and their cancellation mechanism in the standard model.

Chapter 13 contains an elementary introduction to the important subject of the extensions of the standard model that include neutrino masses. We describe the see-saw mechanism, and we present a simple description of the phenomena related to lepton flavour oscillations (see [12] for a review on this subject.)

References

1. C. Itzykson, J.B. Zuber, *Quantum Field Theory* (Mc Graw-Hill, New York, 1980)
2. M.E. Peskin, D.V. Schroeder, *An Introduction to Quantum Field Theory* (Addison Wesley, Redwood City, 1995)
3. S. Weinberg, *The Quantum Theory of Fields* (Cambridge University Press, Cambridge, 1995)
4. R.K. Ellis, W.J. Stirling, B.R. Webber, *QCD and Collider Physics* (Cambridge University Press, Cambridge 1996)
5. L.B. Okun, *Leptons and Quarks* (North-Holland, Amsterdam, 1982)
6. G. Altarelli, M.W. Grunewald, Phys. Rept. **403–404**, 189 (2004). [hep-ph/0404165]
7. G. Altarelli, R. Barbieri, F. Caravaglios, Int. J. Mod. Phys. A **13**, 1031 (1998). [hep-ph/9712368]
8. C.T. Sachrajda, in Menstrup 1997, High-energy physics, pp. 150–189. [hep-ph/9801343]
9. A.J. Buras, in *Theory and experiment heading for new physics*, Erice (2000). [hep-ph/0101336]
10. J.R. Ellis, M.K. Gaillard, D.V. Nanopoulos, Nucl. Phys. B **106**, 292 (1976)
11. M.A. Shifman, A.I. Vainshtein, M.B. Voloshin, V.I. Zakharov, Sov. J. Nucl. Phys. **30**, 711 (1979). [Yad. Fiz. **30** (1979) 1368]
12. M.C. Gonzalez-Garcia, Y. Nir, Rev. Mod. Phys. **75**, 345 (2003). [hep-ph/0202058]

Chapter 2
Relativistic Field Theory

2.1 Scalar Fields

The dynamical variables that describe relativistic systems are *fields*, that is, functions defined in each point of ordinary space. Important examples are the electromagnetic fields, and Dirac and Yukawa fields. The field description of a physical system opens the way to a direct implementation of the principle of covariance, that guarantees the invariance of the equations of motion under changes of reference frame, and of the principle of causality, which is connected to the principle of locality, namely, the independence of variables associated to different points in space at the same time.

The equations of motion for a field theory follow from the principle of stationarity of the action functional, which is defined, on the basis of covariance and locality, as an integral over the three-dimensional space and over the relevant time interval, of a Lagrangian density:

$$S = \int_{t_1}^{t_2} dt \int d^3r\, \mathcal{L}(t, \boldsymbol{r}). \tag{2.1}$$

The Lagrangian density $\mathcal{L}$ is assumed to be a local function of the fields and their first derivatives, so that the equations of motion contain at most second derivatives.

Lorentz invariance of the action functional is achieved if $\mathcal{L}$ transforms as a scalar field: going from a reference frame O to O' through a Lorentz transformation $x' = \Lambda x$, such as for example

$$\begin{pmatrix} ct' \\ x' \\ y' \\ z' \end{pmatrix} = \begin{pmatrix} \frac{1}{\sqrt{1-\beta^2}} & 0 & 0 & \frac{\beta}{\sqrt{1-\beta^2}} \\ 0 & \cos\theta & \sin\theta & 0 \\ 0 & -\sin\theta & \cos\theta & 0 \\ \frac{\beta}{\sqrt{1-\beta^2}} & 0 & 0 & \frac{1}{\sqrt{1-\beta^2}} \end{pmatrix} \begin{pmatrix} ct \\ x \\ y \\ z \end{pmatrix} \tag{2.2}$$

C. M. Becchi and G. Ridolfi, *An Introduction to Relativistic Processes and the Standard Model of Electroweak Interactions*, UNITEXT for Physics,
DOI: 10.1007/978-3-319-06130-6_2, © Springer International Publishing Switzerland 2014

the Lagrangian density must transform as

$$\mathcal{L}'(x) = \mathcal{L}(\Lambda^{-1}x). \tag{2.3}$$

We begin by considering the simple case of a single real field ϕ, which is assumed to transform as a scalar under Lorentz transformations. By analogy with classical mechanics, the Lagrangian density has the form

$$\mathcal{L} = \frac{1}{2c^2}\left(\frac{\partial \phi}{\partial t}\right)^2 - \frac{1}{2}(\nabla\phi)^2 - V(\phi), \tag{2.4}$$

where $V(\phi)$, the *scalar potential*, is independent of the field derivatives, and bounded from below. (With this choice, the field ϕ has the dimension of the square root of an energy divided by a length.) We recognize the typical structure of a Lagrangian, difference between a kinetic term and a potential term. A fundamental constraint, called *renormalizability*, requires that $V(\phi)$ be a polynomial in ϕ of degree not larger than four; this point will be discussed in more detail in Sect. 4.5.

Equation (2.4) can be cast in a manifestly covariant form:

$$\mathcal{L} = \frac{1}{2}\,\partial^\mu\phi\,\partial_\mu\phi - V(\phi), \tag{2.5}$$

where ∂^μ is the four-vector formed with the partial derivatives with respect to space-time coordinates.[1]

Let us now consider the variation of the action functional under a generic infinitesimal variation $\delta\phi$ of ϕ, localized in space, with the conditions $\delta\phi(t_1, \boldsymbol{r}) = \delta\phi(t_2, \boldsymbol{r}) = 0$. We find

$$\begin{aligned}\delta S &= \int_{t_1}^{t_2} dt \int d^3r \left[\frac{\partial \mathcal{L}}{\partial \phi}\,\delta\phi + \frac{\partial \mathcal{L}}{\partial \partial^\mu \phi}\,\delta\partial^\mu\phi\right] \\ &= \int_{t_1}^{t_2} dt \int d^3r \left[\frac{\partial \mathcal{L}}{\partial \phi} - \partial^\mu \frac{\partial \mathcal{L}}{\partial \partial^\mu \phi}\right]\delta\phi. \end{aligned} \tag{2.6}$$

Requiring stability of S under such field variations, and exploiting the arbitrariness of $\delta\phi$, we get the field equations

$$\frac{\partial \mathcal{L}}{\partial \phi} - \partial^\mu \frac{\partial \mathcal{L}}{\partial \partial^\mu \phi} = 0. \tag{2.7}$$

---

[1] Throughout these lectures, we adopt the convention $a^\mu a_\mu = a_0^2 - |\boldsymbol{a}|^2$.

In the case of the Lagrangian density Eq. (2.5) we find

$$\partial^2\phi + \frac{dV(\phi)}{d\phi} = 0. \tag{2.8}$$

If $V(\phi)$ is quadratic in the field ϕ, Eq. (2.8) becomes linear, and can be made homogeneous by a field translation:

$$\partial^2\phi + a\phi = 0. \tag{2.9}$$

We shall see later in this chapter that, after quantization, this equation describes the propagation of non-interacting relativistic particles.

From the point of view of classical field theory we show in Appendix A that a generic solution of Eq. (2.9) vanishes, for large t, as $t^{-3/2}$. As a consequence of this behaviour, if $V(\phi)$ has its absolute minimum in $\phi = 0$, one can show that the large-time solutions of Eq. (2.8) approach those of the corresponding linear equation for sufficiently small initial values of the field and its derivatives. In other words, the asymptotically vanishing solutions become attractors, since the non-linear terms in the field equation vanish faster than the linear terms.

In the context of high energy physics, one studies scattering processes of relativistic particles, that involve asymptotically free particle behaviour. This implies a selection of potentials leading to asymptotically free, and hence asymptotically linear, field equations. Therefore in the case of scalar fields we are led to require the potential to have its absolute minimum in $\phi = 0$. The simplest non-trivial example of such choice is

$$V(\phi) = \frac{1}{2}\mu^2\phi^2 + \frac{\lambda}{4!\hbar c}\phi^4 \tag{2.10}$$

with λ positive. The corresponding free equation is

$$\partial^2\phi + \mu^2\phi = 0. \tag{2.11}$$

2.2 Symmetries in Field Theory

Let us consider an infinitesimal deformation of the field:

$$\phi \to \phi + \alpha\,\delta\phi, \tag{2.12}$$

where α is a constant infinitesimal parameter. We will call this a *global symmetry* transformation if it leaves unaltered the equations of motion, that is, if under (2.12) the action S is unchanged, or equivalently if

$$\mathcal{L} \to \mathcal{L} + \alpha \partial^\mu K_\mu, \tag{2.13}$$

where K_μ is some function of x. In this case, we have

$$\begin{aligned}\delta\mathcal{L} &= \frac{\partial \mathcal{L}}{\partial \phi}\alpha\delta\phi + \frac{\partial \mathcal{L}}{\partial \partial^\mu \phi}\alpha\partial^\mu\delta\phi \\ &= \alpha\delta\phi\left[\frac{\partial \mathcal{L}}{\partial \phi} - \partial^\mu \frac{\partial \mathcal{L}}{\partial \partial^\mu \phi}\right] + \alpha\partial^\mu\left[\frac{\partial \mathcal{L}}{\partial \partial^\mu \phi}\delta\phi\right] \\ &= \alpha\partial^\mu K_\mu.\end{aligned} \tag{2.14}$$

Using the equations of motion Eq. (2.7), we find that the vector current

$$J^\mu = \frac{\partial \mathcal{L}}{\partial \partial_\mu \phi}\delta\phi - K^\mu \tag{2.15}$$

obeys the continuity equation

$$\partial^\mu J_\mu = 0. \tag{2.16}$$

If the field ϕ vanishes sufficiently fast at space infinity, one can define a charge

$$Q(t) = \int d^3r\, J^0(t, \boldsymbol{r}), \tag{2.17}$$

and Eq. (2.16) implies

$$\frac{d}{dt}Q(t) = 0 \tag{2.18}$$

as a consequence of the symmetry property of the Lagrangian density Eq. (2.13).

The correspondence between conserved currents and invariance under continuous symmetries, that reduce to transformations like Eq. (2.12) for infinitesimal parameters, is guaranteed by a general theorem, originally proved by E. Noether.

As an example, let us consider constant space-time translations:

$$x^\mu \to x^\mu + a^\mu \tag{2.19}$$

$$\delta\phi = a^\mu \partial_\mu \phi. \tag{2.20}$$

If the Lagrangian density does not depend explicitly on x, $\mathcal{L} = \mathcal{L}\,(\phi(x), \partial\phi(x))$, then

$$\delta\mathcal{L} = \frac{\partial \mathcal{L}}{\partial \phi}a^\mu\partial_\mu\phi + \frac{\partial \mathcal{L}}{\partial \partial_\nu \phi}a^\mu\partial^\nu\partial_\mu\phi = a^\mu\partial^\nu\left[\frac{\partial \mathcal{L}}{\partial \partial^\nu \phi}\partial_\mu\phi\right] \tag{2.21}$$

by the equations of motion. On the other hand,

$$\delta\mathcal{L} = a^\mu\partial_\mu\mathcal{L} = a^\mu\partial^\nu(g_{\mu\nu}\mathcal{L}), \tag{2.22}$$

and therefore

$$\partial^\mu T_{\mu\nu} = 0 \tag{2.23}$$

where

$$T_{\mu\nu} = \frac{\partial \mathcal{L}}{\partial \partial^\mu \phi} \partial_\nu \phi - g_{\mu\nu} \mathcal{L} \tag{2.24}$$

is called the *energy-momentum tensor*, because the physical quantities that are conserved as a consequence of space-time translation invariance are the total energy and momentum. For example, the total energy of the system is given by

$$E = \int d^3r \, T_{00} = \int d^3r \left(\frac{\partial \mathcal{L}}{\partial \partial^0 \phi} \partial_0 \phi - \mathcal{L} \right). \tag{2.25}$$

In the case of the field theory defined by (2.10) this gives

$$U = \frac{1}{2c^2} \left(\frac{\partial \phi}{\partial t} \right)^2 + \frac{1}{2} \left[(\nabla \phi)^2 + \mu^2 \phi^2 \right] + \frac{\lambda}{4! \hbar c} \phi^4. \tag{2.26}$$

2.3 Particle Interpretation

The scalar field theory outlined in Sect. 2.1 leads naturally to a particle interpretation. To show this, we consider the energy density in the free limit:

$$U_0 = \frac{1}{2c^2} \left(\frac{\partial \phi}{\partial t} \right)^2 + \frac{1}{2} \left[(\nabla \phi)^2 + \mu^2 \phi^2 \right]. \tag{2.27}$$

We restrict the system to a finite volume V of the three-dimensional space, imposing suitable boundary conditions of the field, and we compute the total energy $E = \int_V d^3r \, U_0$ in terms of the Fourier modes of the field:

$$\phi(t, \boldsymbol{r}) = \frac{1}{\sqrt{V}} \sum_{\boldsymbol{k}} \tilde{\phi}_{\boldsymbol{k}}(t) \, e^{i \boldsymbol{k} \cdot \boldsymbol{r}}. \tag{2.28}$$

We obtain

$$E = \frac{1}{2c^2} \sum_{\boldsymbol{k}} \left[|\partial_t \tilde{\phi}_{\boldsymbol{k}}|^2 + \omega_{\boldsymbol{k}}^2 |\tilde{\phi}_{\boldsymbol{k}}|^2 \right], \tag{2.29}$$

where

$$\frac{\omega_{\boldsymbol{k}}^2}{c^2} = |\boldsymbol{k}|^2 + \mu^2. \tag{2.30}$$

Equation (2.29) has the same form as the energy of a system of simple harmonic oscillators, one for each value of $\boldsymbol{k}$. Quantization is therefore straightforward: we define creation and annihilation operators in term of $\tilde{\phi}_{\boldsymbol{k}}$ and $\partial_t \tilde{\phi}_{\boldsymbol{k}}$, now interpreted as operators in the Schrödinger picture, as

$$\tilde{\phi}_{\boldsymbol{k}} = c\sqrt{\frac{\hbar}{2\omega_{\boldsymbol{k}}}}\left(A_{\boldsymbol{k}} + A^{\dagger}_{-\boldsymbol{k}}\right), \qquad \partial_t \tilde{\phi}_{\boldsymbol{k}} = -ic\sqrt{\frac{\hbar\omega_{\boldsymbol{k}}}{2}}\left(A_{\boldsymbol{k}} - A^{\dagger}_{-\boldsymbol{k}}\right). \tag{2.31}$$

and we impose the commutation rules

$$\left[A_{\boldsymbol{k}}, A^{\dagger}_{\boldsymbol{q}}\right] = \delta_{\boldsymbol{k},\boldsymbol{q}}. \tag{2.32}$$

The quantum states of the system are obtained operating with $A^{\dagger}_{\boldsymbol{k}}$ on a vacuum state $|0\rangle$. From the expression of the total energy we obtain the Hamiltonian

$$H = \sum_{\boldsymbol{k}} \hbar\omega_{\boldsymbol{k}}\, A^{\dagger}_{\boldsymbol{k}} A_{\boldsymbol{k}} + \text{constant}. \tag{2.33}$$

The constant term must be set to zero, in order that the vacuum state be Lorentz-invariant. Using the results of Sect. 2.2, one can compute the total momentum components

$$P_i = \frac{1}{c}\int d^3r\, T_{0i} \tag{2.34}$$

in terms of creation and annihilation operators. The result is

$$\boldsymbol{P} = \sum_{\boldsymbol{k}} \hbar\boldsymbol{k}\, A^{\dagger}_{\boldsymbol{k}} A_{\boldsymbol{k}}. \tag{2.35}$$

Clearly, the states $A^{\dagger}_{\boldsymbol{k}}|0\rangle$ are eigenvectors of both H and $\boldsymbol{P}$ with eigenvalues

$$\hbar\omega_{\boldsymbol{k}} = \sqrt{c^2\hbar^2(\boldsymbol{k}^2 + \mu^2)} \tag{2.36}$$

and

$$\hbar\boldsymbol{k} \tag{2.37}$$

respectively. It is therefore natural to interpret them as states of free particles with momentum $\hbar\boldsymbol{k}$ and mass $m = \hbar\frac{\mu}{c}$. This is the reason why the quadratic part of the Lagrangian density is called the free part. The non-quadratic interaction terms will be collectively denoted by $\mathcal{L}_I$.

We denote with $|\{N_{\boldsymbol{q}}\}\rangle$ the state with $N_{\boldsymbol{q}}$ particles with momentum $\hbar\boldsymbol{q}$ for each value of $\boldsymbol{q}$. The commutation rules (2.32) give

$$A_{\boldsymbol{k}}|\{N_{\boldsymbol{q}}\}\rangle = \sqrt{N_{\boldsymbol{k}}}\,|\{N_{\boldsymbol{q}} - \delta_{\boldsymbol{k},\boldsymbol{q}}\}\rangle \tag{2.38}$$

$$A^{\dagger}_{\boldsymbol{k}}|\{N_{\boldsymbol{q}}\}\rangle = \sqrt{N_{\boldsymbol{k}} + 1}\,|\{N_{\boldsymbol{q}} + \delta_{\boldsymbol{k},\boldsymbol{q}}\}\rangle \tag{2.39}$$

and therefore

$$A^{\dagger}_{\boldsymbol{k}}A_{\boldsymbol{k}}|\{N_{\boldsymbol{q}}\}\rangle = N_{\boldsymbol{k}}|\{N_{\boldsymbol{q}}\}\rangle. \tag{2.40}$$

Notice that the particles described by the real scalar field obey Bose statistics: each state can be occupied by an arbitrary number of particles, and the wave functions are symmetric under coordinate permutations. Fermi statistics will appear in Sect. 5.1, where fields describing spin-1/2 particles are introduced.

The expression of the field in the Heisenberg picture evolved at time t is immediately obtained from Eq. (2.28):

$$\phi(t,\boldsymbol{r}) = c\sqrt{\frac{\hbar}{V}}\sum_{\boldsymbol{k}}\frac{e^{i\boldsymbol{k}\cdot\boldsymbol{r}}}{\sqrt{2\omega_{\boldsymbol{k}}}}\left(A_{\boldsymbol{k}}e^{-i\omega_{\boldsymbol{k}}t} + A^{\dagger}_{-\boldsymbol{k}}e^{i\omega_{\boldsymbol{k}}t}\right). \tag{2.41}$$

In the infinite-volume limit the discrete variable $\boldsymbol{k}$ takes values in the continuum, and sums over $\boldsymbol{k}$ must be replaced by three-dimensional integrals:

$$\sum_{\boldsymbol{k}} \to \frac{V}{(2\pi)^3}\int d^3k. \tag{2.42}$$

Correspondingly,

$$\delta_{\boldsymbol{k},\boldsymbol{q}} \to \frac{(2\pi)^3}{V}\,\delta(\boldsymbol{k}-\boldsymbol{q}). \tag{2.43}$$

We now define operators $A(\boldsymbol{k})$, functions of the continuum variable $\boldsymbol{k}$, as

$$A(\boldsymbol{k}) = \sqrt{\frac{V}{(2\pi)^3}}\,A_{\boldsymbol{k}}, \tag{2.44}$$

so that

$$\left[A(\boldsymbol{k}), A^{\dagger}(\boldsymbol{q})\right] = \delta(\boldsymbol{k}-\boldsymbol{q}). \tag{2.45}$$

Hence,

$$\phi(t,\boldsymbol{r}) = c\sqrt{\frac{\hbar}{V}}\sum_{\boldsymbol{k}}\frac{e^{i\boldsymbol{k}\cdot\boldsymbol{r}}}{\sqrt{2\omega_{\boldsymbol{k}}}}\left(A_{\boldsymbol{k}}e^{-i\omega_{\boldsymbol{k}}t} + A^{\dagger}_{-\boldsymbol{k}}e^{i\omega_{\boldsymbol{k}}t}\right)$$

$$= c\sqrt{\frac{\hbar}{V}\frac{V}{(2\pi)^3}}\int d^3k\sqrt{\frac{(2\pi)^3}{V}}\frac{e^{i\boldsymbol{k}\cdot\boldsymbol{r}}}{\sqrt{2\omega_{\boldsymbol{k}}}}\left[A(\boldsymbol{k})e^{-i\omega_{\boldsymbol{k}}t}+A^\dagger(-\boldsymbol{k})e^{i\omega_{\boldsymbol{k}}t}\right]$$

$$= \frac{c\sqrt{\hbar}}{(2\pi)^{3/2}}\int d^3k\,\frac{e^{i\boldsymbol{k}\cdot\boldsymbol{r}}}{\sqrt{2\omega_{\boldsymbol{k}}}}\left[A(\boldsymbol{k})\,e^{-i\omega_{\boldsymbol{k}}t}+A^\dagger(-\boldsymbol{k})\,e^{i\omega_{\boldsymbol{k}}t}\right]. \tag{2.46}$$

Notice that in this formula the variable $\boldsymbol{k}$ is a wave number and $\omega_{\boldsymbol{k}}$ is an angular frequency. A more physical expression is obtained replacing the variable $\boldsymbol{k}$ by the corresponding momentum $\boldsymbol{p} = \hbar\boldsymbol{k}$, $\omega_{\boldsymbol{k}}$ by $\frac{E_p}{\hbar}$ and $A(\boldsymbol{k})$ by $\hbar^{\frac{3}{2}}A(\boldsymbol{p})$, where $A^\dagger(\boldsymbol{p})$ is the creation operator of a single particle state with momentum $\boldsymbol{p}$ and normalization $\langle\boldsymbol{p}|\boldsymbol{p}'\rangle = \delta(\boldsymbol{p}-\boldsymbol{p}')$. In this way one gets

$$\phi(t,\boldsymbol{r}) = \frac{c}{\sqrt{(2\pi)^3\hbar}}\int\frac{d\boldsymbol{p}}{\sqrt{2E_p}}[A(\boldsymbol{p})e^{-i\frac{p\cdot x}{\hbar}}+A^\dagger(\boldsymbol{p})e^{i\frac{p\cdot x}{\hbar}}] \tag{2.47}$$

where

$$p\cdot x = E_p t - \boldsymbol{p}\cdot\boldsymbol{r}. \tag{2.48}$$

Furthermore, it is worth noticing that Eqs. (2.45) and (2.47) are equivalent to the canonical equal-time commutation rule

$$[\phi(t,\boldsymbol{r}),\dot{\phi}(t,\boldsymbol{r}')] = ic^2\hbar\delta(\boldsymbol{r}-\boldsymbol{r}'). \tag{2.49}$$

2.4 Complex Scalar Fields and Antiparticles

Let us now consider a system described by two real scalar fields ϕ_1, ϕ_2. The Lagrangian density is a function of the fields and their derivatives, $\mathcal{L}(\phi_i, \partial\phi_i)$, and the condition of stationarity of S leads to a system of field equations:

$$\partial_\mu\frac{\partial\mathcal{L}}{\partial\partial_\mu\phi_i}-\frac{\partial\mathcal{L}}{\partial\phi_i}=0\quad,\quad i=1,2. \tag{2.50}$$

Let us assume that the Lagrangian density is invariant under the global transformation

$$\begin{pmatrix}\phi_1'(x)\\ \phi_2'(x)\end{pmatrix} = \begin{pmatrix}\cos\alpha & \sin\alpha\\ -\sin\alpha & \cos\alpha\end{pmatrix}\begin{pmatrix}\phi_1(x)\\ \phi_2(x)\end{pmatrix}. \tag{2.51}$$

To first order in α, Eq. (2.51) becomes

$$\phi_i'(x)-\phi_i(x) = \alpha\sum_{j=1}^{2}\epsilon_{ij}\phi_j(x), \tag{2.52}$$

where ϵ_{ij} is the antisymmetric tensor in two dimensions:

$$\epsilon_{12} = 1, \qquad \epsilon_{21} = -1, \qquad \epsilon_{11} = \epsilon_{22} = 0. \tag{2.53}$$

This transformation is a rotation in the two-dimensional space whose points are identified by the 'coordinates' ϕ_1, ϕ_2; this will be called the *isotopic space*, and has obviously nothing to do with the physical space. The assumed invariance of the Lagrangian density implies

$$\sum_{i,j=1}^{2} \left[\partial_\mu \left(\epsilon_{ij} \alpha \phi_j(x) \right) \frac{\partial \mathcal{L}(x)}{\partial \partial_\mu \phi_i(x)} + \epsilon_{ij} \alpha \phi_j(x) \frac{\partial \mathcal{L}(x)}{\partial \phi_i(x)} \right] = 0 \tag{2.54}$$

(assuming $K^\mu = 0$.) Since α is arbitrary and x-independent, this becomes

$$\begin{aligned} &\sum_{i,j=1}^{2} \epsilon_{ij} \left[\partial_\mu \phi_j(x) \frac{\partial \mathcal{L}(x)}{\partial \partial_\mu \phi_i(x)} + \phi_j(x) \frac{\partial \mathcal{L}(x)}{\partial \phi_i(x)} \right] \\ &\quad = \sum_{i,j=1}^{2} \epsilon_{ij} \left[\partial_\mu \phi_j(x) \frac{\partial \mathcal{L}(x)}{\partial \partial_\mu \phi_i(x)} + \phi_j(x) \partial_\mu \frac{\partial \mathcal{L}}{\partial \partial_\mu \phi_i} \right] \\ &\quad = \partial_\mu \left[\sum_{i,j=1}^{2} \epsilon_{ij} \phi_j(x) \frac{\partial \mathcal{L}(x)}{\partial \partial_\mu \phi_i(x)} \right] = 0, \end{aligned} \tag{2.55}$$

where we have used Eq. (2.50). Hence

$$\partial_\mu J^\mu(x) = 0; \qquad J^\mu(x) = \sum_{i,j=1}^{2} \epsilon_{ij} \phi_j(x) \frac{\partial \mathcal{L}(x)}{\partial \partial_\mu \phi_i(x)}. \tag{2.56}$$

The most general renormalizable Lagrangian density invariant under the transformation Eq. (2.51) is

$$\mathcal{L} = \frac{1}{2} \left[(\partial \phi_1)^2 + (\partial \phi_2)^2 - \mu^2 \left(\phi_1^2 + \phi_2^2 \right) \right] - \frac{\lambda}{4\hbar c} \left(\phi_1^2 + \phi_2^2 \right)^2, \tag{2.57}$$

which gives

$$J_\mu = \phi_2 \partial_\mu \phi_1 - \phi_1 \partial_\mu \phi_2. \tag{2.58}$$

Notations simplify considerably if we introduce a complex scalar field

$$\Phi(x) = \frac{1}{\sqrt{2}} \left[\phi_1(x) + i \phi_2(x) \right]. \tag{2.59}$$

The transformation in Eq. (2.51) reduces to the multiplication of the complex field by a phase factor:

$$\Phi'(x) = e^{i\alpha}\Phi(x). \tag{2.60}$$

Furthermore, we find

$$\mathcal{L} = \partial\Phi^* \, \partial\Phi - \mu^2\Phi^*\Phi - \frac{\lambda}{\hbar c}\left(\Phi^*\Phi\right)^2, \tag{2.61}$$

and the conserved current takes the form

$$J_\mu = i\left[\Phi^*\partial_\mu\Phi - \Phi\partial_\mu\Phi^*\right]. \tag{2.62}$$

The quantization of the complex scalar field Φ is an obvious generalization of the real field case. Since the equation of motion for the free field is the same as for the real scalar field, the corresponding free-field solution is also of the same kind, with the only difference that in this case the coefficients of the plane waves with positive and negative frequency are not hermitian conjugates of each other, because the field is complex:

$$\Phi(t, \boldsymbol{r}) = \frac{c}{\sqrt{(2\pi)^3\hbar}} \int \frac{d^3p}{\sqrt{2E_{\boldsymbol{p}}}} \left[A(\boldsymbol{p})\, e^{-i\frac{p\cdot x}{\hbar}} + B^\dagger(\boldsymbol{p})\, e^{i\frac{p\cdot x}{\hbar}}\right]. \tag{2.63}$$

Thus, we have now particles of two different species, A-particles created by the operators $A^\dagger(\boldsymbol{k})$, and B-particles created by the operators $B^\dagger(\boldsymbol{k})$, with equal masses.

The continuity relation $\partial_\mu J^\mu = 0$ corresponds to the conservation of the charge

$$Q = \int d^3r\, J^0(x) = i\int d^3r\left[\Phi^*\partial^0\Phi - \Phi\partial^0\Phi^*\right]. \tag{2.64}$$

A straightforward calculation yields

$$Q = \int d^3k\left[A^\dagger(\boldsymbol{k})A(\boldsymbol{k}) - B^\dagger(\boldsymbol{k})B(\boldsymbol{k})\right], \tag{2.65}$$

where an additive constant was fixed by the requirement

$$Q|0\rangle = 0. \tag{2.66}$$

Recalling Eq. (2.40), we see that Q is just the number of A-particles minus the number of B-particles. Any additive quantum number carried by A-particles, such as electric charge, baryon or lepton number, strangeness etc., is therefore conserved, provided B-particles carry the same quantum number, with opposite sign. If e is the electric charge carried by A particles, the current eJ may be interpreted as the electromagnetic four-current associated to the complex field Φ, which is therefore to be interpreted as a charged scalar field.

We conclude that, for each charged particle (say A-particles in our example), the theory predicts the existence of another charged particle with equal mass and opposite charge (B particles in our case), which is usually called the *antiparticle*. This is part of a very general result, called the CPT theorem, that establishes the correspondence between matter and antimatter.

Chapter 3
Scattering Theory

3.1 Cross Sections

Before discussing scattering processes in field theory, we briefly recall a few general results in scattering theory in the traditional Lippman-Schwinger formulation.

The purpose of scattering theory is the description of asymptotic-time transitions between different states of a system of particles that undergo short-range interactions. In other words, the typical situation is the transition from a state in the remote past composed of particles largely separated in space, and therefore non-interacting, to another state with similar features in the far future. It is natural to think of the full Hamiltonian H, that describes the evolution of the system at all times, as the sum of an asymptotic Hamiltonian H_0, that describes the asymptotic evolution of non-interacting particles, and an interaction term V. Both H and H_0 are assumed to have the same spectrum, which, for simplicity, is assumed to consist of an isolated value, associated with the vacuum state, and a purely continuum component.

In the Lippman-Schwinger formulation, an asymptotic state is in correspondence to a generic state $|\varphi_{\boldsymbol{k},n}\rangle$ of n non-interacting particles with definite momenta $\boldsymbol{k}_i$, $i = 1, \ldots, n$. Here we shall limit our discussion to the case of a system of identical spin-0 bosons, where

$$|\varphi_{\boldsymbol{k},n}\rangle = \prod_{i=1}^{n} A^\dagger(\boldsymbol{k}_i)|0\rangle. \tag{3.1}$$

The states $|\varphi_{\boldsymbol{k},n}\rangle$ are assumed to form a generalized orthonormal basis in the space of free-particle states, that is, of generalized eigenstates of H_0. Next, one builds two distinct bases of generalized eigenstates of H by

$$|\Psi^\pm_{\boldsymbol{k},n}\rangle = |\varphi_{\boldsymbol{k},n}\rangle + \frac{1}{E_{\boldsymbol{k}} - H_0 \pm i\epsilon} V\, |\Psi^\pm_{\boldsymbol{k},n}\rangle, \tag{3.2}$$

C. M. Becchi and G. Ridolfi, *An Introduction to Relativistic Processes and the Standard Model of Electroweak Interactions*, UNITEXT for Physics,
DOI: 10.1007/978-3-319-06130-6_3, © Springer International Publishing Switzerland 2014

where ϵ is a positive infinitesimal, and

$$E_{\boldsymbol{k}} = E_{\boldsymbol{k}_1} + \cdots + E_{\boldsymbol{k}_n}. \tag{3.3}$$

We assume, for simplicity, that no bound state is present in the spectrum. We further assume asymptotic completeness, that is, that the system of states $|\Psi^{\pm}_{\boldsymbol{k},n}\rangle$, together with the vacuum state $|\Omega\rangle$ of the interacting theory, form two equivalent bases of the same Fock space. Thus,

$$|\Psi^{\pm}_{\boldsymbol{k},n}\rangle = \prod_{i=1}^{n} A^{\dagger}_{\text{in/out}}(\boldsymbol{k}_i)|\Omega\rangle, \tag{3.4}$$

where the operators $A^{\dagger}_{\text{in/out}}(\boldsymbol{k}_i)$ are asymptotic particle creation operators.

The states $|\Psi^{\pm}_{\boldsymbol{k},n}\rangle$ are usually referred to as *incoming* (+) and *outgoing* (−) states. This nomencature arises from the following considerations. Let us build an m-particle wave packet out of the states $|\Psi^{+}_{\boldsymbol{k},m}\rangle$ as

$$|\Psi^{+}_{g}\rangle = \int D^{3m}Q\, g(\boldsymbol{Q})\, |\Psi^{+}_{\boldsymbol{Q},m}\rangle, \tag{3.5}$$

where we have introduced the shorthand notation

$$D^{3m}Q = d^3Q_1 \dots d^3Q_m \tag{3.6}$$

$$g(\boldsymbol{Q}) = \psi_{\boldsymbol{p}_1}(\boldsymbol{Q}_1) \dots \psi_{\boldsymbol{p}_m}(\boldsymbol{Q}_m). \tag{3.7}$$

Here $\psi_{\boldsymbol{p}_0}(\boldsymbol{p})$ is a wave packet with average momentum $\boldsymbol{p}_0$, which can be chosen for example to have the Gaussian shape[1]

$$\psi_{\boldsymbol{p}_0}(\boldsymbol{p}) = \frac{1}{(\sqrt{\pi}\delta)^{3/2}}\, e^{-\frac{(\boldsymbol{p}-\boldsymbol{p}_0)^2}{2\delta^2}}; \qquad \int d^3p\, |\psi_{\boldsymbol{p}_0}(\boldsymbol{p})|^2 = 1. \tag{3.8}$$

Note that, for $\delta \to 0$,

$$\psi_{\boldsymbol{p}_0}(\boldsymbol{p}) \underset{\delta\to 0}{\to} N_\delta \delta(\boldsymbol{p}-\boldsymbol{p}_0); \qquad N_\delta = \int d^3p\, \psi_{\boldsymbol{p}_0}(\boldsymbol{p}) = \left(\sqrt{4\pi}\delta\right)^{\frac{3}{2}}. \tag{3.9}$$

Using the identity between distributions

$$\int_{\mp\infty}^{0} d\tau\, e^{-ix\tau} = \frac{i}{x \pm i\epsilon}, \tag{3.10}$$

[1] We shall systematically use the Gaussian shape since this highly simplifies the calculations without any physically relevant loss of generality.

Equation (3.2) gives

$$|\Psi_g^+\rangle = \int D^{3m}Q\, g(\boldsymbol{Q}) \left[|\varphi_{\boldsymbol{Q},m}\rangle - \frac{i}{\hbar} \int_{-\infty}^{0} d\tau e^{-\frac{i}{\hbar}(E_Q - H_0)\tau} V |\Psi_{\boldsymbol{Q},m}^+\rangle \right]. \quad (3.11)$$

The time evolution of this wave packet is therefore

$$\begin{aligned} |\Psi_g^+(t)\rangle &= \int D^{3m}Q\, g(\boldsymbol{Q})\, e^{-\frac{i}{\hbar}E_Q t}\, |\Psi_{\boldsymbol{Q},m}^+\rangle \\ &= \int D^{3m}Q\, g(\boldsymbol{Q})\, e^{-\frac{i}{\hbar}H_0 t} \left[|\varphi_{\boldsymbol{Q},m}\rangle - \frac{i}{\hbar} \int_{-\infty}^{t} d\tau\, e^{-\frac{i}{\hbar}(E_Q - H_0)\tau}\, V |\Psi_{\boldsymbol{Q},m}^+\rangle \right]. \end{aligned} \quad (3.12)$$

In the limit $t \to -\infty$ the wave packet $|\Psi_g^+(t)\rangle$ approaches the analogous wave packet built with the non-interacting states $|\varphi_{\boldsymbol{Q},m}\rangle$

$$|\varphi_g\rangle = \int D^{3m}Q\, g(\boldsymbol{Q}) |\varphi_{\boldsymbol{Q},m}\rangle, \quad (3.13)$$

thereby explaining its identification with an incoming state. Similarly,

$$\begin{aligned} |\Psi_f^-(t)\rangle &= \int D^{3n}q f(\boldsymbol{q})\, e^{-\frac{i}{\hbar}E_q t}\, |\Psi_{\boldsymbol{q},n}^-\rangle \\ &= \int D^{3n}q f(\boldsymbol{q})\, e^{-\frac{i}{\hbar}H_0 t} \left(|\varphi_{\boldsymbol{q},n}\rangle - \frac{i}{\hbar} \int_{+\infty}^{t} d\tau\, e^{-\frac{i}{\hbar}(E_q - H_0)\tau}\, V |\Psi_{\boldsymbol{q},n}^-\rangle \right), \end{aligned} \quad (3.14)$$

where

$$f(\boldsymbol{q}) = \psi_{\boldsymbol{k}_1}(\boldsymbol{q}_1) \ldots \psi_{\boldsymbol{k}_n}(\boldsymbol{q}_n), \quad (3.15)$$

approaches the analogous wave packet built with the non-interacting states

$$|\varphi_f\rangle = \int D^{3n}q f(\boldsymbol{q}) |\varphi_{\boldsymbol{q},n}\rangle \quad (3.16)$$

in the far future $t \to +\infty$, and therefore is identified with an outgoing state. Note that the asymptotic limits $t \to \pm\infty$ can only be taken on sufficiently smooth wave packets, since the time dependence of the scattering states $|\Psi_{\boldsymbol{k},n}^{\pm}(t)\rangle$ is oscillatory. This is usually called the weak topology limit in the Hilbert space.

The two different bases of the state space $|\Psi_{\boldsymbol{q},n}^-\rangle$, $|\Psi_{\boldsymbol{Q},m}^+\rangle$ are related by a unitary transformation; the unitary matrix that relates the coefficients of the decomposition of any state vector in the two bases is called the *scattering matrix*, or simply $\mathcal{S}$ matrix, and its elements are given by

$$\mathcal{S}_{qQ}^{nm} = \langle \Psi_{q,n}^{-} | \Psi_{Q,m}^{+} \rangle . \tag{3.17}$$

Any state vector $|\Psi\rangle$ can be decomposed on either bases. Using Eq. (3.2) one can show that

$$\mathcal{S}_{qQ}^{nm} = \mathcal{I}_{qQ}^{nm} - 2\pi i \delta(E_q - E_Q) \mathcal{T}_{qQ}^{nm} , \tag{3.18}$$

where

$$\mathcal{I}_{qQ}^{nm} = \delta_{nm} \delta(\boldsymbol{q}_1 - \boldsymbol{Q}_1) \ldots \delta(\boldsymbol{q}_n - \boldsymbol{Q}_n) + \text{permutations} \tag{3.19}$$

and

$$\begin{aligned} -2\pi i\, \delta\left(E_q - E_Q\right) \mathcal{T}_{qQ}^{nm} &= -2\pi i\, \delta\left(E_q - E_Q\right) \langle \varphi_{q,n} | V | \Psi_{Q,m}^{+} \rangle \\ &= -2\pi i\, \delta\left(E_q - E_Q\right) \langle \Psi_{q,n}^{-} | V | \varphi_{Q,m} \rangle . \end{aligned} \tag{3.20}$$

A proof of Eqs. (3.18) and (3.20) is given in Appendix B.

The unitarity of $\mathcal{S}$, that follows from asymptotic completeness, is expressed by

$$\sum_{\ell} \frac{1}{\ell!} \int D^{3\ell} k\, \mathcal{S}_{qk}^{n\ell} \left(\mathcal{S}_{Qk}^{m\ell} \right)^{*} = \mathcal{I}_{qQ}^{nm} , \tag{3.21}$$

where a factor of $1/\ell!$ was inserted to take into account that particles in the intermediate states are identical. In the general case when the intermediate states contain ℓ_s identical particles of species s, this factor must be replaced by $1/\prod_s \ell_s!$.

Replacing Eq. (3.18) in Eq. (3.21), we obtain the unitaritycondition

$$\begin{aligned} &\delta(E_q - E_Q) \left[\mathcal{T}_{qQ}^{nm} - \left(\mathcal{T}_{Qq}^{mn} \right)^{*} \right] \\ &\quad = -2\pi i \delta(E_q - E_Q) \sum_{\ell} \frac{1}{\ell!} \int D^{3\ell} k\, \delta(E_q - E_k) \mathcal{T}_{qk}^{n\ell} \left(\mathcal{T}_{Qk}^{m\ell} \right)^{*} . \end{aligned} \tag{3.22}$$

Some important consequences of the constraint Eq. (3.22), and the relationship between unitarity and renormalizability, will be briefly discussed in Sect. 4.5.

In the following, we will restrict our attention to transition processes in which the initial state is substantially different from the final state: even when the number and species of particles are the same in both states, we will require that the particle momenta are all different from each other, and that no subset of final-state particles has a total momentum equal to that of any subset of initial state particles. Under these assumptions, the term $\mathcal{I}_{qQ}^{nm}$ in Eq. (3.18) is identically zero. The quantity

$$\mathcal{A}^{nm}_{\boldsymbol{kp}} = \int D^{3n}qD^{3m}Q f^*(\boldsymbol{q})g(\boldsymbol{Q})\mathcal{S}^{nm}_{\boldsymbol{qQ}}$$
$$= -2\pi i \int D^{3n}qD^{3m}Q f^*(\boldsymbol{q})g(\boldsymbol{Q})\delta\left(E_{\boldsymbol{q}} - E_{\boldsymbol{Q}}\right) \mathcal{T}^{nm}_{\boldsymbol{qQ}} \tag{3.23}$$

is called the *scattering amplitude* for the transition from an initial state of m particles characterized by the wave packet $g(\boldsymbol{Q})$, and a final state of n particles characterized by the wave packet $f(\boldsymbol{q})$. We further restrict ourselves to the case of two particles in the initial state ($m = 2$), that is

$$f(\boldsymbol{q}) = \psi_{\boldsymbol{k}_1}(\boldsymbol{q}_1) \ldots \psi_{\boldsymbol{k}_n}(\boldsymbol{q}_n) \tag{3.24}$$
$$g(\boldsymbol{Q}) = \psi_{\boldsymbol{p}_1}(\boldsymbol{Q}_1)\, \psi_{\boldsymbol{p}_2}(\boldsymbol{Q}_2). \tag{3.25}$$

In the cases of interest, invariance under space translations implies momentum conservation, and we may write

$$\mathcal{T}^{n2}_{\boldsymbol{qQ}} = \delta(\boldsymbol{q} - \boldsymbol{Q})\, T(\boldsymbol{q}, \boldsymbol{Q}), \tag{3.26}$$

where

$$\boldsymbol{q} = \boldsymbol{q}_1 + \cdots + \boldsymbol{q}_n; \qquad \boldsymbol{Q} = \boldsymbol{Q}_1 + \boldsymbol{Q}_2. \tag{3.27}$$

In such cases

$$\mathcal{S}^{n2}_{\boldsymbol{qQ}} = -2\pi i \delta(q - Q) T(\boldsymbol{q}, \boldsymbol{Q}) \tag{3.28}$$

and

$$\mathcal{A}^{n2}_{\boldsymbol{kp}} = -2\pi i \int D^{3n}q\, D^6Q\, \delta(q - Q) f^*(\boldsymbol{q}) T(\boldsymbol{q}, \boldsymbol{Q}) g(\boldsymbol{Q}) \tag{3.29}$$

where

$$\delta(q - Q) = \delta\left(E_{\boldsymbol{q}} - E_{\boldsymbol{Q}}\right)\, \delta(\boldsymbol{q} - \boldsymbol{Q}). \tag{3.30}$$

In the typical scattering experiment, two colliding beams, described by initial wave packets, are prepared by means of suitable acceleration and collimation devices. The final state is then analyzed by a system of detectors. In many (but not all) physically relevant cases one of the two initial beams is at rest in the laboratory reference frame; we will however discuss the scattering process in full generality. The initial-state wave packets are determined by the momentum and energy resolutions of the accelerator, and by the geometric parameters that define its luminosity. On the other hand, the final state selection is limited to a certain domain in the space of final momenta, corresponding to the acceptance of the detectors. Therefore, what one is really interested in is the production probability of a given number of final particles in a given region of their momentum space.

The final goal of the present analysis is the final state production probability density, which will be determined starting form the amplitude $\mathcal{A}^{n2}_{\boldsymbol{kp}}$ for high final wave packets momentum resolution. For this reason, we will keep the wave packet resolution δ_i of initial-state particles different from the resolution δ_f of particles in the final state, and we will consider the limit $\delta_f \to 0$. As a second step, we will take the limit $\delta_i \to 0$ in the sense of distributions, which is appropriate in the case of a probability density.

A crucial ingredient in the calculation of the probability density is the assumption that $T(\boldsymbol{q}, \boldsymbol{Q})$ be, at least locally, a uniformly continuous function, and hence essentially a constant over the integration range defined by the wave packets in Eq. (3.29). Under this assumption, Eq. (3.29) gives

$$\left|\mathcal{A}^{n2}_{\boldsymbol{kp}}\right|^2 \simeq 4\pi^2 \left|T(\boldsymbol{k},\boldsymbol{p})\right|^2 \int D^6QD^6Q'D^{3n}qD^{3n}q' \\ \delta(Q-q)\delta(Q'-q')f^*(\boldsymbol{q})g(\boldsymbol{Q})f(\boldsymbol{q}')g^*(\boldsymbol{Q}'). \tag{3.31}$$

We now define the production probability density in the momentum space of the final particles. To this purpose, we define the density matrix $\rho_F(\boldsymbol{q},\boldsymbol{q}')$ in the final particle momenta by

$$\left|\mathcal{A}^{n2}_{\boldsymbol{kp}}\right|^2 = \int D^{3n}qD^{3n}q' \rho_F(\boldsymbol{q},\boldsymbol{q}')f^*(\boldsymbol{q})f(\boldsymbol{q}'). \tag{3.32}$$

In the limit $\delta_f \to 0$,

$$f(\boldsymbol{q}) \to (2\sqrt{\pi}\delta_f)^{\frac{3n}{2}}\delta(\boldsymbol{q}_1-\boldsymbol{k}_1)\ldots\delta(\boldsymbol{q}_n-\boldsymbol{k}_n) \tag{3.33}$$

and therefore, in the same limit,

$$\left|\mathcal{A}^{n2}_{\boldsymbol{kp}}\right|^2 = (2\sqrt{\pi}\delta_f)^{3n}\rho_F(\boldsymbol{k},\boldsymbol{k}). \tag{3.34}$$

Comparing Eqs. (3.31) and (3.34) we find

$$\begin{aligned}\rho_F(\boldsymbol{k},\boldsymbol{k}) &= (2\pi)^2\left|T(\boldsymbol{k},\boldsymbol{p})\right|^2 \int D^6QD^6Q'\, g(\boldsymbol{Q})g^*(\boldsymbol{Q}')\delta(Q-k)\delta(Q'-k) \\ &= (2\pi)^2(\sqrt{\pi}\delta_i)^{-6}\left|T(\boldsymbol{k},\boldsymbol{p})\right|^2 \int D^6QD^6Q'\,\delta(Q-Q')\delta(Q-k) \\ &\quad e^{-\frac{1}{2\delta_i^2}\sum_{i=1}^2(Q_i-p_i)^2}e^{-\frac{1}{2\delta_i^2}\sum_{i=1}^2(Q'_i-p_i)^2}.\end{aligned} \tag{3.35}$$

The integral is most easily performed in terms of the integration variables X_i, Y_i defined by

$$Q_i = p_i + \delta_i\left(X_i + \frac{Y_i}{2}\right), \qquad Q'_i = p_i + \delta_i\left(X_i - \frac{Y_i}{2}\right) \tag{3.36}$$

We get

$$\rho_F(\boldsymbol{k},\boldsymbol{k}) = \frac{4\delta_i^6}{\pi}\left|T(\boldsymbol{k},\boldsymbol{p})\right|^2 I_X(\delta_i)I_Y(\delta_i), \tag{3.37}$$

where

$$I_X(\delta_i) = \int D^6X\, \delta(p - k + \delta_i X) e^{-\sum_{i=1}^{2} X_i^2} \tag{3.38}$$

$$I_Y(\delta_i) = \int D^6Y\, \delta(\delta_i Y) e^{-\frac{1}{4}\sum_{i=1}^{2} Y_i^2} \tag{3.39}$$

and

$$p = \sum_{i=1}^{2} p_i, \quad k = \sum_{j=1}^{n} k_j, \quad X = \sum_{i=1}^{2} X_i, \quad Y = \sum_{i=1}^{2} Y_i \tag{3.40}$$

The two integrals $I_X(\delta_i)$ and $I_Y(\delta_i)$ are easily computed in the small-δ_i limit. We find

$$\begin{aligned} I_Y(\delta_i) &\to \frac{1}{\delta_i^4} \int d^3Y_1 d^3Y_2\, e^{-\frac{1}{4}(\boldsymbol{Y}_1^2+\boldsymbol{Y}_2^2)} \delta^3(\boldsymbol{Y}_1 + \boldsymbol{Y}_2) \delta(\boldsymbol{V}_1 \cdot \boldsymbol{Y}_1 + \boldsymbol{V}_2 \cdot \boldsymbol{Y}_2) \\ &= \frac{2\pi}{\delta_i^4 |\boldsymbol{V}_1 - \boldsymbol{V}_2|}, \end{aligned} \tag{3.41}$$

where

$$\boldsymbol{V}_i = \frac{\boldsymbol{p}_i}{E_{\boldsymbol{p}_i}} = \frac{\boldsymbol{p}_i}{\sqrt{|\boldsymbol{p}_i|^2 + m_i^2}} \tag{3.42}$$

are the initial particle velocities. Also,

$$\begin{aligned} I_X(\delta_i) &\to \int d^3X_1 d^3X_2\, e^{-\boldsymbol{X}_1^2 - \boldsymbol{X}_2^2} \delta^3(\boldsymbol{p} - \boldsymbol{k} + \delta_i(\boldsymbol{X}_1 + \boldsymbol{X}_2)) \\ &\quad \delta(E_{\boldsymbol{p}} - E_{\boldsymbol{k}} + \delta_i(\boldsymbol{V}_1 \cdot \boldsymbol{X}_1 + \boldsymbol{V}_2 \cdot \boldsymbol{X}_2)) \\ &= \frac{\pi}{2\delta_i^4 |\boldsymbol{V}_1 - \boldsymbol{V}_2|} e^{-\frac{(\boldsymbol{p}-\boldsymbol{k})^2}{2\delta_i^2}} e^{-\frac{2}{\delta_i^2 |\boldsymbol{V}_1 - \boldsymbol{V}_2|^2}\left[E_{\boldsymbol{p}} - E_{\boldsymbol{k}} - \frac{1}{2}(\boldsymbol{p}-\boldsymbol{k})\cdot(\boldsymbol{V}_1 + \boldsymbol{V}_2)\right]^2}. \end{aligned} \tag{3.43}$$

Equations (3.41, 3.43) show that the probability, considered as a limited function, that is in the Banach topology, vanishes in the limit $\delta_i = \delta_f = \delta \to 0$ as δ^{3n-2}:

$$\left| \mathcal{A}_{\boldsymbol{kp}}^{n2} \right|^2 \sim \delta^{3n-2}. \tag{3.44}$$

However we also have

$$\lim_{\delta_i \to 0} I_X(\delta_i) = \pi^3 \delta(p - k) \tag{3.45}$$

in the sense of distributions, and hence, in the same sense,

$$\rho_F(\boldsymbol{k}, \boldsymbol{k}) \to (2\pi)^3 \delta_i^2 \frac{|T(\boldsymbol{k}, \boldsymbol{p})|^2}{|\boldsymbol{V}_1 - \boldsymbol{V}_2|} \delta(p - k). \tag{3.46}$$

We now turn to a discussion of the physical interpretation of our result, Eq. (3.46). In all cases of interest, the initial-state wave packets do not spread significantly during the scattering process; hence, the time evolution of the corresponding probability densities

$$\rho_i(t, \boldsymbol{r}) = |\psi_i(t, \boldsymbol{r})|^2 = \left| \frac{1}{(2\pi\hbar)^{\frac{3}{2}}} \int d^3Q\, \psi_{\boldsymbol{p}_i}(\boldsymbol{Q}) e^{\frac{i}{\hbar}\boldsymbol{Q}\cdot\boldsymbol{r}} e^{-\frac{i}{\hbar}E_Q t} \right|^2 \tag{3.47}$$

is simply a rigid translation:

$$\rho_i(t, \boldsymbol{r}) = \rho_i(\boldsymbol{r} - \boldsymbol{V}_i t), \qquad i = 1, 2\,. \tag{3.48}$$

We define the *integrated luminosity* L of the prepared system as

$$L = \lim_{d \to 0} \frac{P(d_{12} < d)}{\pi d^2}, \tag{3.49}$$

where $P(d_{12} < d)$ is the probability that, in the absence of interactions, the two colliding particles come to a relative distance less than d.[2]

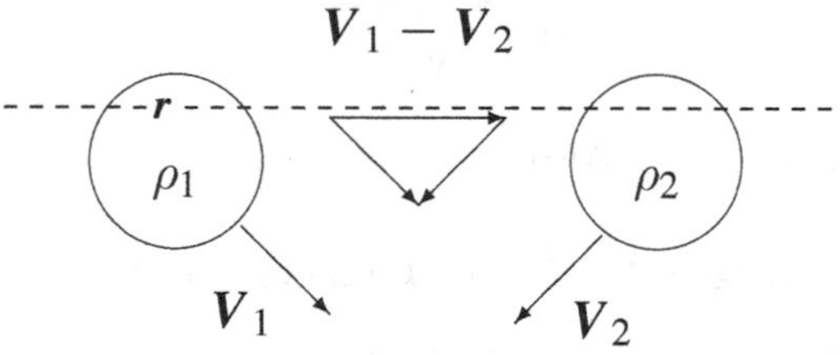

The integrated luminosity is easily computed in terms of the probability densities ρ_1, ρ_2 of the initial state particles. Consider the above picture: the probability that a particle of beam 1, initially at $\boldsymbol{r}$, come closer than d to one of the particles of beam 2 is given, in the small-d limit, by πd^2 times the integral of ρ_2 along the dashed line:

$$\pi d^2 \int_{-\infty}^{+\infty} dt\, |\boldsymbol{V}_1 - \boldsymbol{V}_2|\, \rho_2(\boldsymbol{r} + (\boldsymbol{V}_1 - \boldsymbol{V}_2)t). \tag{3.50}$$

[2] This definition of integrated luminosity is the same as the one currently adopted in experimental particle physics, referred to the case of a single collision.

The probability $P(d_{12} < d)$ is obtained by integrating this quantity over all possible choices of $\boldsymbol{r}$ with weight $\rho_1(\boldsymbol{r})$:

$$\begin{aligned} L &= |\boldsymbol{V}_1 - \boldsymbol{V}_2| \int d^3r\, dt\, \rho_1(\boldsymbol{r})\, \rho_2(\boldsymbol{r} + (\boldsymbol{V}_1 - \boldsymbol{V}_2)t) \\ &= |\boldsymbol{V}_1 - \boldsymbol{V}_2| \int d^3r\, dt\, \rho_1(\boldsymbol{r} - \boldsymbol{V}_1 t)\, \rho_2(\boldsymbol{r} - \boldsymbol{V}_2 t). \end{aligned} \tag{3.51}$$

Using Eq. (3.47) we find

$$\begin{aligned} L &= \frac{|\boldsymbol{V}_1 - \boldsymbol{V}_2|}{(2\pi\hbar)^6} \int d^3r\, dt \int D^6Q D^6Q'\, g(\boldsymbol{Q}) g^*(\boldsymbol{Q}') e^{\frac{i}{\hbar}(\boldsymbol{Q}-\boldsymbol{Q}')\cdot \boldsymbol{r}}\, e^{-\frac{i}{\hbar}(E_Q - E_{Q'})t} \\ &= \frac{|\boldsymbol{V}_1 - \boldsymbol{V}_2|}{(2\pi\hbar)^2} \int D^6Q D^6Q'\, g^*(\boldsymbol{Q}') g(\boldsymbol{Q}) \delta(Q - Q'). \end{aligned} \tag{3.52}$$

The integrated luminosity for two Gaussian packets, Eq. (3.8), with average momenta $\boldsymbol{p}_i$, $i = 1, 2$, in the limit $\delta_i \to 0$ can be computed following the same procedure that led us to the computation of the squared amplitude, Eq. (3.31). We find

$$L_{\delta_i} = \frac{\delta_i^2}{2\pi\hbar^2}. \tag{3.53}$$

Thus,

$$\rho_F(\boldsymbol{k}, \boldsymbol{k}) = \hbar^2 L_{\delta_i} \frac{|T(\boldsymbol{k}, \boldsymbol{p})|^2}{|\boldsymbol{V}_1 - \boldsymbol{V}_2|} (2\pi)^4 \delta(p - k). \tag{3.54}$$

This "exact" factorization property of the final density into the product of the integrated luminosity and a δ_i-independent distribution, up to corrections of order δ_i^2, appears as a direct consequence of the use of Gaussian packets. However, its validity is in fact completely general in the limit of high resolution packets. Indeed, it is apparent from Eq. (3.35) that $\rho_F(\boldsymbol{k}, \boldsymbol{k})$, as a function of the total final-state four-momentum k, has a sharp peak around the total initial-state four-momentum p, whose width vanishes in the high resolution limit independently of the shape of the wave packets:

$$\rho_F(\boldsymbol{k}, \boldsymbol{k}) \simeq C\delta(p - k), \tag{3.55}$$

with

$$C = \int d^4k\, \rho_F(\boldsymbol{k}, \boldsymbol{k}) = (2\pi)^2 |T(\boldsymbol{k}, \boldsymbol{p})|^2 \int D^6Q D^6Q' \delta(Q - Q') g(\boldsymbol{Q}) g^*(\boldsymbol{Q}'). \tag{3.56}$$

From Eq. (3.52), one sees that the integral factor is equal to the integrated luminosity divided by $\frac{|\boldsymbol{V}_1 - \boldsymbol{V}_2|}{(2\pi\hbar)^2}$.

Now, we define the *differential cross section* $d\sigma$ through

$$\rho_F(\boldsymbol{k},\boldsymbol{k})\,d\boldsymbol{k}_1\ldots d\boldsymbol{k}_n = L\,d\sigma, \tag{3.57}$$

or

$$d\sigma = \frac{(2\pi)^4\hbar^2}{|\boldsymbol{V}_1-\boldsymbol{V}_2|}|T(\boldsymbol{k},\boldsymbol{p})|^2\,\delta(p-k)\,d\boldsymbol{k}_1\ldots.d\boldsymbol{k}_n, \tag{3.58}$$

where, we recall, we have defined

$$p = p_1 + p_2; \qquad k = k_1 + \cdots + k_n. \tag{3.59}$$

The amplitude $\mathcal{A}^{n2}_{\boldsymbol{kp}}$ is in a sense an elementary transiton amplitude. It is related to the physical quantities in the δ_i, $\delta_f \to 0$ limit through the identity

$$\left|\mathcal{A}^{n2}_{\boldsymbol{kp}}\right|^2 = L_{\delta_i}\frac{d\sigma}{d\Gamma}\Gamma_{\delta_f}, \tag{3.60}$$

where L_{δ_i} is the integrated luminosity defined in Eq. (3.52), which vanishes as δ_i^2 in the limit $\delta_i \to 0$. The factor Γ_{δ_f} is the final state phase space delimited by final-state wave packets; it is proportional to δ_f^{3n-4}. Finally, $d\sigma/d\Gamma$ is the δ-independent differential cross section for the process.

3.2 Decay Rates

A remarkable amount of information on the standard model comes from the study of the decay properties of unstable particles. The best known example is the width of the Z^0 boson, which, for example, provides information on the number of light neutrinos. In field theory, unstable particles play a delicate role since, strictly speaking, they are not true particles, in the sense that they do not appear in the space of scattering asymptotic states. The reason for this is that radiative corrections introduce a complex correction to the mass of an unstable particle, whose imaginary part is related to the particle decay probability per unit time. This point will be discussed in Sect. 12.4. In most cases of interest, one considers a single-particle state

$$|\Psi_i\rangle = \int d^3p\,g\,(\boldsymbol{p})\,|\Psi_{\boldsymbol{p},1}\rangle. \tag{3.61}$$

The particle is assumed to be stable in the absence of the small perturbation H_W, which induces the decay. In such case, one can use the Fermi golden rule to obtain the decay probability per unit time from the initial state $|\Psi_i\rangle$ to a final scattering state $|\Psi_f^-\rangle$:

$$\frac{dP(t)}{dt} = \frac{2\pi}{\hbar}\left|\langle\Psi_f^-|H_W|\Psi_i\rangle\right|^2\rho(E_i), \tag{3.62}$$

where $\rho(E)$ is the density of final states with energy E. This formula is a direct consequence of first order, time-dependent perturbation theory.

The final state is a scattering state, Eq. (3.2), of n particles with momenta $\boldsymbol{k}_1, \ldots, \boldsymbol{k}_n$. As in the case of elastic scattering, one has

$$\langle \Psi^{-}_{\boldsymbol{k},n} | H_W | \Psi_{\boldsymbol{p},1} \rangle = \delta(\boldsymbol{p} - \boldsymbol{k}) T(\boldsymbol{k}, \boldsymbol{p}) \,, \tag{3.63}$$

with $\boldsymbol{k} = \boldsymbol{k}_1 + \cdots + \boldsymbol{k}_n$, as a consequence of translation invariance. If the final states are selected in a region Δ of the momentum space, Eq. (3.62) becomes

$$\frac{dP_\Delta(t)}{dt} = \frac{2\pi}{\hbar} \int d^3p \, |g(\boldsymbol{p})|^2 \int_\Delta D^{3n}k \, |T(\boldsymbol{k}, \boldsymbol{p})|^2 \, \delta(\boldsymbol{p} - \boldsymbol{k}) \delta(E_p - E_k), \tag{3.64}$$

where E_p is the energy of the initial particle with momentum $\boldsymbol{p}$. Under the same assumptions leading to Eq. (3.58) for the differential cross section, we can define a differential decay probability per unit time as

$$d\Gamma = \frac{2\pi}{\hbar} d^3k_1 \ldots d^3k_n \, |T(\boldsymbol{k}, \boldsymbol{p})|^2 \, \delta(p - k). \tag{3.65}$$

Equation (3.64) is completely analogous to the expression Eq. (3.58) for the differential cross section in the case of elastic scattering.

3.3 Semi-Classical Approximation and Asymptotic Conditions

We have seen in the previous sections that the computation of cross sections and decay rates requires, according to Eqs. (3.58) and (3.65), the evaluation of the amplitude $T(\boldsymbol{k}, \boldsymbol{p})$. In this section, we will perform this calculation in the semi-classical approximation of field theory.

It is well known from ordinary quantum mechanics that probability amplitudes in the semi-classical limit $\hbar \to 0$ are given by

$$\psi = e^{\frac{i}{\hbar}(S + O(\hbar))} \,, \tag{3.66}$$

where S is the Hamilton-Jacobi action function. The simplest example is provided by a single free, non-relativistic particle moving from $\boldsymbol{r}_i$ to $\boldsymbol{r}_f$ in the time interval $t_f - t_i$; the classical motion takes place with velocity $\boldsymbol{v} = \frac{\boldsymbol{r}_f - \boldsymbol{r}_i}{t_f - t_i}$, and the action is

$$S = \frac{1}{2} m \boldsymbol{v}^2 \, (t_f - t_i) = \frac{m(\boldsymbol{r}_f - \boldsymbol{r}_i)^2}{2(t_f - t_i)} \,. \tag{3.67}$$

The corresponding amplitude in the semi-classical limit is therefore

$$\psi = \langle \boldsymbol{r}_f | e^{-\frac{iH(t_f - t_i)}{\hbar}} | \boldsymbol{r}_i \rangle = \exp \frac{i}{\hbar} \left[\frac{m(\boldsymbol{r}_f - \boldsymbol{r}_i)^2}{2(t_f - t_i)} + O(\hbar) \right], \tag{3.68}$$

which gives rise to the familiar interference phenomenon.

We now consider the case of a scattering process characterized by relativistic energies E. The duration of the full process, from the preparation of the initial state to the detection of final state particles, is essentially infinite compared to the characteristic time scale of the interaction, of order $\frac{\hbar}{E}$.

It is shown in Appendix D that, in the limit of infinite momentum resolution of the wave packets, $\delta_i = \delta_f = \delta \to 0$, the amplitude Eq. (3.29) can be written in terms of the (non normalized) coherent states

$$|I\rangle \equiv e^{\sqrt{2}A_I^\dagger} |\Omega\rangle, \qquad |F\rangle \equiv e^{\sqrt{n}A_F^\dagger} |\Omega\rangle \tag{3.69}$$

where

$$A_I^\dagger \equiv \frac{1}{\sqrt{2}} \sum_{j=1}^{2} A_{j,\mathrm{in}}^\dagger, \qquad A_F^\dagger \equiv \frac{1}{\sqrt{n}} \sum_{i=1}^{n} A_{i,\mathrm{out}}^\dagger, \tag{3.70}$$

as

$$\mathcal{A}_{\boldsymbol{kp}}^{n2} = \langle F|I\rangle - 1. \tag{3.71}$$

Following the above example, in the semi-classical approximation the scalar product $\langle F|I\rangle$ should be written in the form

$$\langle F|I\rangle = \exp\left(\frac{iS_{i\to f}}{\hbar}\right) = 1 + \mathcal{A}_{\boldsymbol{kp}}^{n2}, \tag{3.72}$$

where $S_{i\to f}$ is the classical field action whose value must be computed on the solution of the classical field equation interpolating between suitable initial field variables computed in the limit $t \to -\infty$ and final ones in the limit $t \to \infty$.

Equation (3.72) can be further simplified recalling that

$$\mathcal{A}_{\boldsymbol{kp}}^{n2} \sim \delta^{\frac{3n}{2}-1} \tag{3.73}$$

from Eq. (3.44). Indeed, Eq. (3.72) can be written in the form

$$S_{i\to f} = -i\hbar \log\left(1 + \mathcal{A}_{\boldsymbol{kp}}^{n2}\right), \tag{3.74}$$

which implies that

$$S_{i\to f} \sim \delta^{\frac{3n}{2}-1} + O(\delta^{3n-2}), \tag{3.75}$$

and hence

$$\mathcal{A}_{\boldsymbol{kp}}^{n2} = e^{\frac{iS_{i\to f}}{\hbar}} - 1 = \frac{iS_{i\to f}}{\hbar}[1 + O(\delta^{\frac{3}{2}n-1})]. \tag{3.76}$$

In the limit of infinite momentum resolution, the scattering amplitude is directly proportional to $S_{i\to f}$.

In order to identify the mentioned field variables it is convenient to take the three-dimensional Fourier transform of the field:

$$\tilde{\phi}(t,\boldsymbol{k}) = \frac{1}{(2\pi)^3}\int d^3r\, e^{-i\boldsymbol{k}\cdot\boldsymbol{r}}\phi(t,\boldsymbol{r}). \tag{3.77}$$

To simplify notations, we choose from now on the so-called natural units, where $\hbar = c = 1$. In these units, we have

$$m = \mu; \qquad E_{\boldsymbol{k}} = \sqrt{|\boldsymbol{k}|^2 + m^2}. \tag{3.78}$$

The Fourier transform of a free field $\phi_0(x)$, Eq. (2.47), is

$$\tilde{\phi}_0(t,\boldsymbol{k}) = \frac{1}{\sqrt{(2\pi)^3 2E_{\boldsymbol{k}}}}[A(\boldsymbol{k})e^{-iE_{\boldsymbol{k}}t} + A^\dagger(-\boldsymbol{k})e^{iE_{\boldsymbol{k}}t}]. \tag{3.79}$$

It is possible to isolate the annihilation or the creation operator in this expression by means of the following identity:

$$i\sqrt{\frac{(2\pi)^3}{2E_{\boldsymbol{k}}}}\left[e^{iE_{\boldsymbol{k}}t}\frac{\partial\tilde{\phi}_0(t,\boldsymbol{k})}{\partial t} - \frac{\partial e^{iE_{\boldsymbol{k}}t}}{\partial t}\tilde{\phi}_0(t,\boldsymbol{k})\right] = A(\boldsymbol{k}). \tag{3.80}$$

The creation operator $A^\dagger(\boldsymbol{k})$ is simply obtained by hermitian conjugation, using $\tilde{\phi}^\dagger(t,\boldsymbol{k}) = \tilde{\phi}(t,-\boldsymbol{k})$. For a generic interacting scalar field configuration $\phi(x)$, the combination in the lhs of Eq. (3.80) is, in general, time-dependent. However, in the weak operator topology, the limits

$$\lim_{t\to\pm\infty} i\sqrt{\frac{(2\pi)^3}{2E_{\boldsymbol{k}}}}\left[e^{iE_{\boldsymbol{k}}t}\frac{\partial\tilde{\phi}(t,\boldsymbol{k})}{\partial t} - \frac{\partial e^{iE_{\boldsymbol{k}}t}}{\partial t}\tilde{\phi}(t,\boldsymbol{k})\right] \tag{3.81}$$

exist, and are proportional to $A_{\text{in}}(\boldsymbol{k})$ and $A_{\text{out}}(\boldsymbol{k})$:

$$\lim_{t\to-\infty} i\sqrt{\frac{(2\pi)^3}{2E_{\boldsymbol{k}}}}\left[e^{iE_{\boldsymbol{k}}t}\frac{\partial\tilde{\phi}(t,\boldsymbol{k})}{\partial t} - \frac{\partial e^{iE_{\boldsymbol{k}}t}}{\partial t}\tilde{\phi}(t,\boldsymbol{k})\right] = \sqrt{Z}A_{\text{in}}(\boldsymbol{k}) \tag{3.82}$$

$$\lim_{t\to+\infty} i\sqrt{\frac{(2\pi)^3}{2E_{\boldsymbol{k}}}}\left[e^{iE_{\boldsymbol{k}}t}\frac{\partial\tilde{\phi}(t,\boldsymbol{k})}{\partial t} - \frac{\partial e^{iE_{\boldsymbol{k}}t}}{\partial t}\tilde{\phi}(t,\boldsymbol{k})\right] = \sqrt{Z}A_{\text{out}}(\boldsymbol{k}). \tag{3.83}$$

In a fully quantized field theory this can be proved, using general properties of the theory such as the cluster property and the positivity of particle energies.

In the classical limit, Eqs. (3.82, 3.83) are just consequences of the free asymptotic behavior of the solutions of the field equations. As shown in Appendix A, the free field vanishes as $|t|^{-\frac{3}{2}}$ for large $|t|$, and hence the interaction vanishes faster than the field itself; thus, the field equation approaches the free field equation in this limit. We will show in Sect. 3.4 that $Z = 1$ in the semi-classical approximation.

We now observe that the coherent states $|I\rangle$, $|F\rangle$ are eigenstates of $A_{\text{in}}(\boldsymbol{k})$ and $A_{\text{out}}(\boldsymbol{k})$ respectively:

$$A_{\text{in}}(\boldsymbol{k})|I\rangle = \sum_{i=1}^{2} \psi_{\boldsymbol{p}_i}(\boldsymbol{k})|I\rangle \tag{3.84}$$

$$A_{\text{out}}(\boldsymbol{k})|F\rangle = \sum_{j=1}^{n} \psi_{\boldsymbol{k}_j}(\boldsymbol{k})|F\rangle. \tag{3.85}$$

We conclude that evaluating $\langle F|I\rangle$ in the semi-classical approximation we have to use Eq. (3.71) and compute the value of the classical action assuming the fields asymptotic conditions:

$$i\sqrt{\frac{(2\pi)^3}{2E_{\boldsymbol{k}}}} \lim_{t\to-\infty} \left[e^{iE_{\boldsymbol{k}}t} \frac{\partial\tilde{\phi}(t,\boldsymbol{k})}{\partial t} - \frac{\partial e^{iE_{\boldsymbol{k}}t}}{\partial t}\tilde{\phi}(t,\boldsymbol{k}) \right] = \sum_{i=1}^{2} \psi_{\boldsymbol{p}_i}(\boldsymbol{k})$$

$$-i\sqrt{\frac{(2\pi)^3}{2E_{\boldsymbol{k}}}} \lim_{t\to+\infty} \left[e^{-iE_{\boldsymbol{k}}t} \frac{\partial\tilde{\phi}(t,\boldsymbol{k})}{\partial t} - \frac{\partial e^{-iE_{\boldsymbol{k}}t}}{\partial t}\tilde{\phi}(t,\boldsymbol{k}) \right] = \sum_{j=1}^{n} \psi^*_{\boldsymbol{k}_j}(-\boldsymbol{k}). \tag{3.86}$$

Both asymptotic conditions can be summarized in a single one; indeed, inserting the free-field solution

$$\tilde{\phi}^{(\text{as})}(t,\boldsymbol{k}) = \frac{1}{\sqrt{(2\pi)^3 2E_{\boldsymbol{k}}}} \left[\sum_{i=1}^{2} \psi_{\boldsymbol{p}_i}(\boldsymbol{k})\, e^{-iE_{\boldsymbol{k}}t} + \sum_{j=1}^{n} \psi^*_{\boldsymbol{k}_j}(-\boldsymbol{k})\, e^{iE_{\boldsymbol{k}}t} \right] \tag{3.87}$$

in Eq. (3.86), one finds the same asymptotic limits. Thus, we will require that the interacting field evolve, in the weak sense, towards the asymptotic field Eq. (3.87). The inverse Fourier transform of Eq. (3.87) is

$$\phi^{(\text{as})}(t,\boldsymbol{x}) = \frac{1}{\sqrt{(2\pi)^3}} \int \frac{d^3k}{\sqrt{2E_{\boldsymbol{k}}}} \left[\sum_{i=1}^{2} \psi_{\boldsymbol{p}_i}(\boldsymbol{k})\, e^{-ik\cdot x} + \sum_{j=1}^{n} \psi^*_{\boldsymbol{k}_j}(\boldsymbol{k})\, e^{ik\cdot x} \right]. \tag{3.88}$$

Using Gaussian wave packets in the limit $\delta \to 0$ the asymptotic conditions become

$$\tilde{\phi}^{(\mathrm{as})}(t,\boldsymbol{k}) = \frac{(\sqrt{4\pi\delta})^{\frac{3}{2}}}{(2\pi)^{\frac{3}{2}}}\left[\sum_{i=1}^{2}\frac{\delta\left(\boldsymbol{k}-\boldsymbol{p}_i\right)}{\sqrt{2E_{\boldsymbol{p}_i}}}e^{-iE_{\boldsymbol{k}}t} + \sum_{j=1}^{n}\frac{\delta\left(\boldsymbol{k}+\boldsymbol{k}_j\right)}{\sqrt{2E_{\boldsymbol{k}_j}}}e^{iE_{\boldsymbol{k}}t}\right]$$

$$\phi^{(\mathrm{as})}(t,\boldsymbol{x}) = \frac{(\sqrt{4\pi\delta})^{\frac{3}{2}}}{(2\pi)^{\frac{3}{2}}}\left[\sum_{i=1}^{2}\frac{e^{-ip_i\cdot x}}{\sqrt{2E_{\boldsymbol{p}_i}}} + \sum_{j=1}^{n}\frac{e^{ik_j\cdot x}}{\sqrt{2E_{\boldsymbol{k}_j}}}\right]. \tag{3.89}$$

This method is employed in Appendix E in a simple case, where the particles interact with an external time-dependent potential.

3.4 Solution of the Field Equation

The next step toward the computation of the scattering amplitude in the semi-classical approximation is the calculation of the solution of the field Eq. (2.8) with the asymptotic conditions Eq. (3.89). We consider for simplicity the scalar potential in Eq. (2.10). For later convenience, we also insert an external source J_e, which is equivalent to the substitution $\mathcal{L} \to \mathcal{L} + \phi J_e$. Equation (2.8) takes the form

$$\left(\partial^2 + m^2\right)\phi = -\frac{\lambda}{3!}\phi^3 + J_e \equiv J, \tag{3.90}$$

or, in terms of Fourier transforms,

$$\left(\partial_t^2 + \boldsymbol{k}^2 + m^2\right)\tilde{\phi}(t,\boldsymbol{k}) = \left(\partial_t^2 + E_{\boldsymbol{k}}^2\right)\tilde{\phi}(t,\boldsymbol{k}) = \tilde{J}(t,\boldsymbol{k}). \tag{3.91}$$

We shall find the requested solution by the method of Green's functions. We define the *Green function* $\tilde{\Delta}(t,\boldsymbol{k})$ as the solution of the inhomogeneous equation

$$\left(\partial_t^2 + E_{\boldsymbol{k}}^2\right)\tilde{\Delta}(t,\boldsymbol{k}) = \delta(t) \tag{3.92}$$

with the conditions

$$\tilde{\Delta}(t,\boldsymbol{k}) \sim e^{iE_{\boldsymbol{k}}t} \quad \text{for} \quad t < 0 \tag{3.93}$$

$$\tilde{\Delta}(t,\boldsymbol{k}) \sim e^{-iE_{\boldsymbol{k}}t} \quad \text{for} \quad t > 0. \tag{3.94}$$

One can check that

$$\tilde{\Delta}(t,\boldsymbol{k}) = \frac{i}{2E_{\boldsymbol{k}}}\left[\theta(t)e^{-iE_{\boldsymbol{k}}t} + \theta(-t)e^{iE_{\boldsymbol{k}}t}\right] \tag{3.95}$$

by an explicit calculation. It is easy to see that

$$\begin{aligned}\tilde{\phi}(t,\boldsymbol{k}) &= \tilde{\phi}^{(\mathrm{as})}(t,\boldsymbol{k}) + \int_{-\infty}^{+\infty} dt'\, \tilde{\Delta}\left(t-t',\boldsymbol{k}\right)\tilde{J}\left(\boldsymbol{k},t'\right) \\ &= \tilde{\phi}^{(\mathrm{as})}(t,\boldsymbol{k}) \\ &\quad + \frac{i}{2E_{\boldsymbol{k}}}\left[e^{-iE_{\boldsymbol{k}}t}\int_{-\infty}^{t} dt'\, e^{iE_{\boldsymbol{k}}t'}\tilde{J}(t',\boldsymbol{k}) + e^{iE_{\boldsymbol{k}}t}\int_{t}^{\infty} dt'\, e^{-iE_{\boldsymbol{k}}t'}\tilde{J}(t',\boldsymbol{k})\right] \end{aligned} \tag{3.96}$$

is a solution of Eq. (3.91), and approaches Eq. (3.89) asymptotically: for example, for $t \to -\infty$ the first term in the squared bracket vanishes, and the second one tends to the time Fourier transform of $\tilde{J}$ times a rapidly oscillating phase factor $e^{iE_{\boldsymbol{k}}t}$, that does not contribute in the weak limit. Similarly, the squared bracket vanishes as $t \to +\infty$, and in both cases the asymptotic limit is $\tilde{\phi}^{(\mathrm{as})}$.

It will be useful to write the solution Eq. (3.96) in an explicitly covariant form. To this purpose, we take its inverse Fourier transform:

$$\begin{aligned}\phi(t,\boldsymbol{r}) &= \phi^{(\mathrm{as})}(t,\boldsymbol{r}) + \int dt' \int d^3k\, e^{i\boldsymbol{k}\cdot\boldsymbol{r}}\tilde{\Delta}(t-t',\boldsymbol{k})\,\frac{1}{(2\pi)^3}\int d^3r'\, e^{-i\boldsymbol{k}\cdot\boldsymbol{r}'}J(\boldsymbol{r}',t') \\ &= \phi^{(\mathrm{as})}(x) + \int d^4x'\, J(x')\,\frac{1}{(2\pi)^3}\int d^3k\, e^{i\boldsymbol{k}\cdot(\boldsymbol{r}-\boldsymbol{r}')}\,\tilde{\Delta}(t-t',\boldsymbol{k}), \end{aligned} \tag{3.97}$$

where $x = (t, \boldsymbol{r})$ and $x' = (t', \boldsymbol{r}')$. Using the explicit expression of the Green function $\tilde{\Delta}(t, \boldsymbol{k})$, Eq. (3.95), and the integral representation of the step function

$$\theta(t) = -\frac{1}{2\pi i}\int_{-\infty}^{+\infty} d\omega\, \frac{e^{-i\omega t}}{\omega + i\epsilon}, \tag{3.98}$$

we find

$$\begin{aligned}&\frac{1}{(2\pi)^3}\int d^3k\, e^{i\boldsymbol{k}\cdot\boldsymbol{r}}\,\tilde{\Delta}(t,\boldsymbol{k}) \\ &\quad = -\int \frac{d^3k}{(2\pi)^4}\, e^{i\boldsymbol{k}\cdot\boldsymbol{r}}\,\frac{1}{2E_{\boldsymbol{k}}}\int_{-\infty}^{+\infty} d\omega \left[\frac{e^{-i(E_{\boldsymbol{k}}+\omega)t}}{\omega + i\epsilon} + \frac{e^{i(E_{\boldsymbol{k}}+\omega)t}}{\omega + i\epsilon}\right] \\ &\quad = \int \frac{d^4k}{(2\pi)^4}\, e^{-ik\cdot x}\,\frac{1}{2E_{\boldsymbol{k}}}\left[\frac{1}{E_{\boldsymbol{k}} - k^0 - i\epsilon} + \frac{1}{E_{\boldsymbol{k}} + k^0 - i\epsilon}\right], \end{aligned} \tag{3.99}$$

where we have defined $k^0 = E_{\boldsymbol{k}} + \omega$ in the first term, $k^0 = -E_{\boldsymbol{k}} - \omega$ in the second one, and $k = (k^0, \boldsymbol{k})$. This gives

$$\frac{1}{(2\pi)^3}\int d^3k\, e^{i\boldsymbol{k}\cdot\boldsymbol{r}}\,\tilde{\Delta}(t,\boldsymbol{k}) = \int \frac{d^4k}{(2\pi)^4}\, e^{-ik\cdot x}\,\frac{1}{m^2 - k^2 - i\epsilon} \equiv \Delta(x) \tag{3.100}$$

(the positive infinitesimal ϵ can be redefined at each step, provided its sign is kept unchanged.) The function $\Delta(x)$ is a Lorentz scalar; hence, the covariant form of Eq. (3.96) is

$$\phi(x) = \phi^{(\mathrm{as})}(x) + \int d^4y\, \Delta(x-y)J(y) \equiv \left(\phi^{(\mathrm{as})} + \Delta \circ J\right)(x), \tag{3.101}$$

with $\phi^{(\mathrm{as})}$ given in Eq. (3.88). Equation (3.101) is an implicit solution of the field equation, since J depends on ϕ; the explicit solution can be found using Eq. (3.101) recursively. In practice, we have transformed the differential equation (2.8) into an integral equation, that embodies the boundary conditions.

Finally, we note that

$$(\partial^2 + m^2)\Delta(x) = \delta(x). \tag{3.102}$$

This relation will be useful in the computation of the Green function for fields with different transformation properties with respect to Lorentz transformations.

We are now in a position to show that, in the semi-classical approximation, the factor Z which appears in Eqs. (3.82, 3.83) is equal to 1. In an interacting theory, Lorentz and space-time translation (Poincaré) invariance imply that

$$\langle \boldsymbol{p}|\tilde{\phi}(t,\boldsymbol{k})|\Omega\rangle = \sqrt{Z}\frac{e^{iE_k t}}{\sqrt{2E_k}(2\pi)^{\frac{3}{2}}}\delta^3(\boldsymbol{p}+\boldsymbol{k}) \tag{3.103}$$

for any single particle state $|\boldsymbol{p}\rangle$. Also,

$$\begin{aligned}\langle \Omega|\tilde{\phi}(t,\boldsymbol{k})\tilde{\phi}(t',\boldsymbol{k}')|\Omega\rangle &= \int d^3p\, \langle \Omega|\tilde{\phi}(t,\boldsymbol{k})|\boldsymbol{p}\rangle\langle \boldsymbol{p}|\tilde{\phi}(t',\boldsymbol{k}')|\Omega\rangle + R(t,t',\boldsymbol{k},\boldsymbol{k}')\\ &= \delta(\boldsymbol{k}+\boldsymbol{k}')\frac{Ze^{iE_k(t-t')}}{2E_k(2\pi)^3} + R(t,t',\boldsymbol{k},\boldsymbol{k}'),\end{aligned} \tag{3.104}$$

where $R(t,t',\boldsymbol{k},\boldsymbol{k}')$ denotes the contribution from multiparticle intermediate states. Again it follows from Poincaré invariance that

$$R(t,t',\boldsymbol{k},\boldsymbol{k}') = \delta(\boldsymbol{k}+\boldsymbol{k}')\int_{2m}^{\infty} d\mu\, \rho(\mu)\frac{e^{i\sqrt{k^2+\mu^2}(t-t')}}{2\sqrt{k^2+\mu^2}(2\pi)^3}, \tag{3.105}$$

with $\rho(\mu) \geq 0$. It is also easy to check that, if the field and its time derivative satisfy canonical commutation relations

$$[\phi(t,\boldsymbol{r}), \dot{\phi}(t,\boldsymbol{r}')] = i\delta(\boldsymbol{r}-\boldsymbol{r}'), \tag{3.106}$$

Equations (3.104, 3.105) imply the sum rule

$$Z + \int_{2m}^{\infty} d\mu\, \rho(\mu) = 1. \tag{3.107}$$

Let us now consider the vacuum expectation value of $\phi(x)$. This is a functional of $\phi^{(\mathrm{as})}(x)$ and $J_e(x)$; in particular, for $J_e = 0$ it is entirely determined by the field equations with the boundary conditions identified by $\phi^{(\mathrm{as})}$, while, for $\phi^{(\mathrm{as})} = 0$, it is a functional of J_e, which vanishes at $J_e = 0$. We have therefore

$$\langle \Omega|\phi(x)|\Omega\rangle = \int d^4y\, \Delta_F(x-y)J_e(y) + O(J_e^2). \tag{3.108}$$

We can use ordinary first-order time-dependent perturbation theory, with a perturbing Hamiltonian $H' = -\int d^3x\, \phi(t, \boldsymbol{x})J_e(t, \boldsymbol{x})$, to obtain

$$\int d^4y\, \Delta_F(x-y)J_e(y) = i\int d^4y\, \theta(x_0 - y_0)\langle\Omega|[\phi(x), \phi(y)]|\Omega\rangle J_e(y) + R(x)_{J_e} \tag{3.109}$$

where $R(x)_{J_e}$ is a suitable inhomogeneous term in the perturbative expansion. This quantity is determined by the condition that $\int dy\, \Delta_F(x-y)J_e(y)$ satisfy the field boundary conditions when $\phi^{(\mathrm{as})} = 0$, namely that it has vanishing positive frequencies when $t \to \infty$ and vanishing negative frequencies when $t \to -\infty$. Under such conditions,

$$R(x)_{J_e} = i\int d^4y\, \langle\Omega|\phi(y)\phi(x)|\Omega\rangle J_e(y), \tag{3.110}$$

and hence

$$\Delta_F(x-y) = \theta(x_0 - y_0)\langle\Omega|\phi(x)\phi(y)|\Omega\rangle + \theta(y_0 - x_0)\langle\Omega|\phi(y)\phi(x)|\Omega\rangle. \tag{3.111}$$

Using Eqs. (3.104, 3.105) it is easy to verify that

$$\Delta_F(x) = \int \frac{d^4k}{(2\pi)^4} e^{-ik\cdot x}\left[\frac{Z}{m^2 - k^2 - i\epsilon} + \int_{2m}^{\infty} d\mu \frac{\rho(\mu)}{\mu^2 - k^2 - i\epsilon}\right]. \tag{3.112}$$

Now it is immediate to compare this result with that obtained in the semiclassical approximation, that is $\Delta \circ J_e$, concluding that in the semiclassical approximation $Z = 1$ and $\rho(\mu)$ vanishes.

3.5 Calculation of the Scattering Amplitude

The calculation of the scattering amplitude involves, as already mentioned, the evaluation of the action integral over an infinite time interval:

$$S_{i\to f} = \int_{-\infty}^{+\infty} dt \int d^3r\, \mathcal{L}(\mathbf{r}, t)\,. \tag{3.113}$$

In view of the oscillating behaviour of the asymptotic solution, the integral is not necessarily convergent; we therefore regularize the integrand by replacing $\phi^{(\text{as})}(x)$ with

$$\phi^{(\text{as},\eta)}(x) \equiv e^{-\eta|t|}\phi^{(\text{as})}(x), \tag{3.114}$$

and we take the limit $\eta \to 0^+$ at the end of the calculation. Clearly, $\phi^{(\text{as},\eta)}$ is no longer a solution of the free-field equation; correspondingly,

$$\phi = \phi^{(\text{as},\eta)} + \Delta \circ J \tag{3.115}$$

is not a stationary field configuration for the original action, but for the modified action obtained from the modified Lagrangian density

$$\mathcal{L} \to \mathcal{L} + \phi(\partial^2 + m^2)\phi^{(\text{as},\eta)}. \tag{3.116}$$

The added term vanishes as $\eta \to 0$, but its integral is not necessarily zero in the same limit.

Taking Eq. (3.116) into account, we must evaluate

$$S^{(\eta)}_{i\to f} = -\int d^4x \left[\frac{1}{2}\phi\left(\partial^2 + m^2\right)\left(\phi - 2\phi^{(\text{as},\eta)}\right) + \frac{\lambda}{4!}\phi^4\right]. \tag{3.117}$$

We obtain

$$\begin{aligned}
S^{(\eta)}_{i\to f} &= -\int d^4x \left[\frac{1}{2}\left(\phi^{(\text{as},\eta)} + \Delta\circ J\right)\left(\partial^2 + m^2\right)\left(\Delta\circ J - \phi^{(\text{as},\eta)}\right)\right.\\
&\qquad \left. + \frac{\lambda}{4!}\left(\phi^{(\text{as},\eta)} + \Delta\circ J\right)^4\right]\\
&= \int d^4x \left[\frac{1}{2}\phi^{(\text{as},\eta)}\left(\partial^2 + m^2\right)\phi^{(\text{as},\eta)} - \frac{1}{2}(\Delta\circ J)\left(\partial^2 + m^2\right)\Delta\circ J\right.\\
&\qquad \left. - \frac{\lambda}{4!}\left(\phi^{(\text{as},\eta)} + \Delta\circ J\right)^4\right]
\end{aligned}$$

$$= \int d^4x \left[\frac{1}{2}\phi^{(\text{as},\eta)} \left(\partial^2 + m^2\right) \phi^{(\text{as},\eta)} - \frac{1}{2} J \, \Delta \circ J \right.$$
$$\left. - \frac{\lambda}{4!} \left(\phi^{(\text{as},\eta)} + \Delta \circ J\right)^4 \right], \qquad (3.118)$$

where we have performed a partial integration in the second step, and we have used $\left(\partial^2 + m^2\right) \Delta \circ J = J$ in the last one. The contribution of the first term in brackets,

$$\frac{1}{2} \int d^4x \, \phi^{(\text{as},\eta)} \left(\partial^2 + m^2\right) \phi^{(\text{as},\eta)}$$
$$= \frac{(2\pi)^3}{2} \int_{-\infty}^{+\infty} dt \int d^3k \, \tilde{\phi}^{(\text{as},\eta)}(t, -\boldsymbol{k}) \left(\partial_t^2 + E_{\boldsymbol{k}}^2\right) \tilde{\phi}^{(\text{as},\eta)}(t, \boldsymbol{k}) \qquad (3.119)$$

vanishes as $\eta \to 0$. Indeed, using

$$\tilde{\phi}^{(\text{as},\eta)}(t, \boldsymbol{k}) = e^{-\eta|t|} \left[\phi_+(\boldsymbol{k}) e^{iE_{\boldsymbol{k}}t} + \phi_-(\boldsymbol{k}) e^{-iE_{\boldsymbol{k}}t}\right], \qquad (3.120)$$

we find

$$\left(\partial_t^2 + E_{\boldsymbol{k}}^2\right) \tilde{\phi}^{(\text{as},\eta)}(t, \boldsymbol{k}) = \left(\eta^2 - 2\eta\delta(t)\right) \tilde{\phi}^{(\text{as},\eta)} s(t, \boldsymbol{k})$$
$$- 2i\eta \, \text{sign}(t) \, E_{\boldsymbol{k}} e^{-\eta|t|} \left(\phi_+(\boldsymbol{k}) e^{iE_{\boldsymbol{k}}t} - \phi_-(\boldsymbol{k}) e^{-iE_{\boldsymbol{k}}t}\right). \qquad (3.121)$$

The term proportional to η^2 does not contribute to Eq. (3.119) as $\eta \to 0$, because the time integration is proportional to $1/\eta$, and the term proportional to $\eta\delta(t)$ vanishes, because the delta function makes the integral convergent. The remaining term is

$$- i\eta(2\pi)^3 \int_{-\infty}^{+\infty} dt \, \text{sign}(t) \, e^{-2\eta|t|} \int d^3k \, E_{\boldsymbol{k}} \left[\phi_+(-\boldsymbol{k}) e^{iE_{\boldsymbol{k}}t} + \phi_-(-\boldsymbol{k}) e^{-iE_{\boldsymbol{k}}t}\right]$$
$$\times \left[\phi_+(\boldsymbol{k}) e^{iE_{\boldsymbol{k}}t} - \phi_-(\boldsymbol{k}) e^{-iE_{\boldsymbol{k}}t}\right] + O(\eta)$$
$$= -i\eta \, (2\pi)^3 \int d^3k \, E_{\boldsymbol{k}} \int_{-\infty}^{+\infty} dt \, \text{sign}(t) \, e^{-2\eta|t|}$$
$$\times \left[e^{2iE_{\boldsymbol{k}}t} \phi_+(-\boldsymbol{k}) \phi_+(\boldsymbol{k}) - e^{-2iE_{\boldsymbol{k}}t} \phi_-(-\boldsymbol{k}) \phi_-(\boldsymbol{k})\right] + O(\eta). \qquad (3.122)$$

The time integral can be performed, with the result

$$2\eta\,(2\pi)^3 \int d^3k\, \frac{E_k^2}{\eta^2 + E_k^2}\left[\phi_+(-\boldsymbol{k})\phi_+(\boldsymbol{k}) + \phi_-(-\boldsymbol{k})\phi_-(\boldsymbol{k})\right] + O(\eta), \tag{3.123}$$

which is also vanishing for $\eta \to 0$. Hence, we obtain in this limit

$$S_{i\to f} = -\int d^4x \left[\frac{1}{2} J\,\Delta \circ J + \frac{\lambda}{4!}\left(\phi^{(\mathrm{as})} + \Delta \circ J\right)^4\right], \tag{3.124}$$

where

$$J = -\frac{\lambda}{3!}\left(\phi^{(\mathrm{as})} + \Delta \circ J\right)^3. \tag{3.125}$$

It can be shown[3] that the case of the $\lambda\phi^4$ interaction the result Eq. (3.124) is a sum of terms proportional to $\lambda^k(\phi^{(\mathrm{as})})^{2k+2}$, $k = 1, 2, \ldots$.

We conclude this Section by giving a generalized version of Eq. (3.124). Indeed, using Eq. (3.101), it can be written as

$$S_{i\to f} = -\int d^4x\, \frac{\lambda}{4!}\phi^4(x) - \frac{1}{2}\int d^4x\, d^4y\, J(x)\,\Delta(x-y)\,J(y). \tag{3.126}$$

This expression readily generalizes to any theory, in the form

$$S_{i\to f} = \int d^4x\, \mathcal{L}_I - \frac{1}{2}\int d^4x\, d^4y\, J(x)\,\Delta(x-y)\,J(y). \tag{3.127}$$

When more fields are involved, ϕ_i, $i = 1, \ldots, N$ (not necessarily scalar fields), there will be a source J_i for each of them, given by

$$J_i = \frac{\partial}{\partial \phi_i}\mathcal{L}_I - \partial_\mu \frac{\partial}{\partial(\partial_\mu \phi_i)}\mathcal{L}_I; \qquad i = 1, \ldots, N, \tag{3.128}$$

as one can see by inspection of the Euler-Lagrange equations. Equation (3.124) can now be rewritten by defining the functional

$$S\left[\phi^{(\mathrm{as})}\right] = -\int d^4x \left[\frac{1}{2} J_i\, \Delta_{ij} \circ J_j - \mathcal{L}_I\right]_{\phi_i = \phi_i^{(\mathrm{as})} + (\Delta \circ J)_i}, \tag{3.129}$$

where we have taken into account the fact that, in the general case, the propagator Δ is a $N \times N$ matrix. This formula is immediately applicable to all cases of physical interest.

---

[3] Expanding the r.h.s. of Eq. (3.125) one can show that J is a sum of terms proportional to $\lambda^k(\phi^{(\mathrm{as})})^{2k+1}$; replacing this expansion in Eq. (3.124) yields the announced result.

3.6 The Asymptotic Field: An Explicit Example

The expression in Eq. (3.88) for the asymptotic field can be checked by an explicit calculation. In this section we will compute the scattering amplitude $A^{n2}_{\boldsymbol{kp}}$ Eq. (3.23) for $2 \to n$ scattering in a scalar theory with interaction Lagrangian

$$\mathcal{L}_I = -\frac{g}{(n+2)!}\phi^{n+2}, \tag{3.130}$$

where g is a constant with mass dimension $n-2$. For simplicity, in this case we will take the limit of vanishing momentum spread since the beginning; wave packets will be therefore replaced by delta functions as in Eq. (3.9). Furthermore, the factor of $[(\sqrt{4\pi\delta})^{3/2}]^{n+2}$ arising in this limit will be omitted. Corrections to the semiclassical approximation are of higher order in g; this follows from the analyses in Sects. 4.3 and 4.5. Therefore the amplitude given in Eq. (3.127) can be compared with that obtained in the Born approximation, with the free theory perturbed by

$$V = \frac{g}{(n+2)!}\int d^3r\, \phi_0^{n+2}(0,\boldsymbol{r}). \tag{3.131}$$

This perturbative result can be obtained by selecting the term linear in V in the right-hand side of Eq. (3.20), that is

$$-2\pi\delta(E_{\boldsymbol{k}} - E_{\boldsymbol{p}})\mathcal{T}^{n2}_{\boldsymbol{kp}} = -2\pi\delta(E_{\boldsymbol{k}} - E_{\boldsymbol{p}})\langle\varphi_{\boldsymbol{k},n}|V|\varphi_{\boldsymbol{p},2}\rangle + O(g^2). \tag{3.132}$$

Using Eq. (3.79) we can write V in terms of free-particle creation and annihilation operators:

$$\begin{aligned} V &= \frac{g}{(n+2)!}(2\pi)^3\int\prod_{l=1}^{n+2}\left[d\boldsymbol{q}_l\,\tilde{\phi}_0\left(0,\boldsymbol{q}_l\right)\right]\delta\left(\sum_{l=1}^{n+2}\boldsymbol{q}_l\right) \\ &= \frac{g}{(2\pi)^{\frac{3n}{2}}(n+2)!}\int\prod_{l=1}^{n+2}\left[\frac{d\boldsymbol{q}_l}{\sqrt{2E_{\boldsymbol{q}_l}}}\left[A(\boldsymbol{q}_l) + A^\dagger(-\boldsymbol{q}_l)\right]\right]\delta\left(\sum_{l=1}^{n+2}\boldsymbol{q}_l\right). \end{aligned} \tag{3.133}$$

We can now compute the scattering amplitude in Born approximation by Wick's theorem:

$$\begin{aligned} &-2\pi\delta(E_{\boldsymbol{k}} - E_{\boldsymbol{p}})\langle\varphi_{\boldsymbol{k},n}|V|\varphi_{\boldsymbol{p},2}\rangle \\ &= -2\pi\delta(E_{\boldsymbol{k}} - E_{\boldsymbol{p}})\frac{g}{(2\pi)^{\frac{3n}{2}}}\int\delta\left(\sum_{l=1}^{n+2}\boldsymbol{q}_l\right)\prod_{i=1}^{2}\frac{d\boldsymbol{q}_i\psi_{\boldsymbol{p}_i}(\boldsymbol{q}_i)}{\sqrt{2E_{\boldsymbol{q}_i}}}\prod_{j=1}^{n}\frac{d\boldsymbol{q}_j\psi^*_{\boldsymbol{k}_j}(-\boldsymbol{q}_j)}{\sqrt{2E_{\boldsymbol{q}_j}}} \end{aligned}$$

$$= -\frac{g}{(2\pi)^{\frac{3n-2}{2}}} \frac{\delta(p_1 + p_2 - \sum_{i=1}^{n} k_i)}{\sqrt{2^{n+2} E_{p_1} E_{p_2} \prod_{j=1}^{n} E_{k_j}}}. \tag{3.134}$$

This result must be compared the first term in the right-hand side of Eq. (3.127), computed with

$$\phi(x) = \phi^{(\mathrm{as})}(x) = \frac{1}{(2\pi)^{\frac{3}{2}}} \left[\sum_{i=1}^{2} \frac{e^{-ip_i \cdot x}}{\sqrt{2E_{p_i}}} + \sum_{j=1}^{n} \frac{e^{ik_j \cdot x}}{\sqrt{2E_{k_j}}} \right]. \tag{3.135}$$

We find

$$\begin{aligned} S_{i\to f} &= -\frac{g}{(n+2)!} \int dx \left(\phi^{(\mathrm{as})}(x) \right)^{n+2} \\ &= -\frac{g}{(2\pi)^{\frac{3(n+2)}{2}} (n+2)!} \int dx \left[\sum_{i=1}^{2} \frac{e^{-ip_i \cdot x}}{\sqrt{2E_{p_i}}} + \sum_{j=1}^{n} \frac{e^{ik_j \cdot x}}{\sqrt{2E_{k_j}}} \right]^{n+2} \end{aligned} \tag{3.136}$$

The integral in the right-hand side of Eq. (3.136) can be decomposed into a sum of $(n+2)^{n+2}$ terms, each giving, after integration, a 4-dimensional delta function in the momenta. For a generic choice of particle momenta, only constrained by momentum conservation, it is apparent that only terms proportional to $\delta(p_1 + p_2 - \sum_{i=1}^{n} k_i)$ contribute. Furthermore the number of these terms is $(n+2)!$, and each of them is proportional to

$$\frac{1}{\sqrt{2^{n+2} E_{p_1} E_{p_2} \prod_{j=1}^{n} E_{k_j}}}. \tag{3.137}$$

Therefore we have

$$S_{i\to f} = -\frac{(2\pi)^4 g}{(2\pi)^{\frac{3(n+2)}{2}}} \frac{\delta(p_1 + p_2 - \sum_{i=1}^{n} k_i)}{\sqrt{2^{n+2} E_{p_1} E_{p_2} \prod_{j=1}^{n} E_{k_j}}} \tag{3.138}$$

which coincides with the amplitude given in Eq. (3.134) thus confirming our choice of the semiclassical asymptotic field.

Therefore we have

Chapter 4
Feynman Diagrams

4.1 The Method of Feynman Diagrams

The calculation of the scattering amplitude for a given process is greatly simplified
by a graphical technique, originally introduced by R. Feynman in the context of
quantum electrodynamics. We define the following symbols:

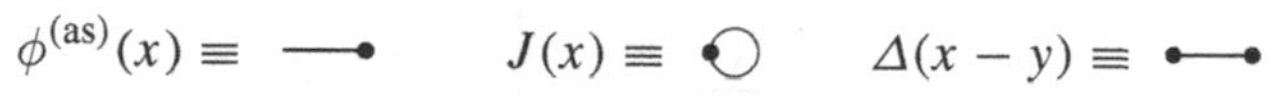

$$\phi^{(as)}(x) \equiv \quad \longrightarrow \bullet \qquad J(x) \equiv \quad \mathbf{O} \qquad \Delta(x - y) \equiv \quad \bullet\!\!\!-\!\!\!\bullet \qquad (4.1)$$

Equation (3.124) is represented as

$$
S_{i \to f} = -\frac{1}{2}\, \mathbf{O}\!\!-\!\!\mathbf{O}
$$

$$
-\frac{\lambda}{4!}\left[\; + \; + 4 \;+\!\!\!\mathbf{O} \; + 6\;\begin{array}{c}\mathbf{O}\\[-2pt]+\\[-2pt]\mathbf{O}\end{array}\; + 4\;\begin{array}{c}\mathbf{O}\\[-2pt]+\!\!\!\mathbf{O}\end{array}\; + \;\mathbf{O}\!+\!\!\mathbf{O}\;\right].
$$

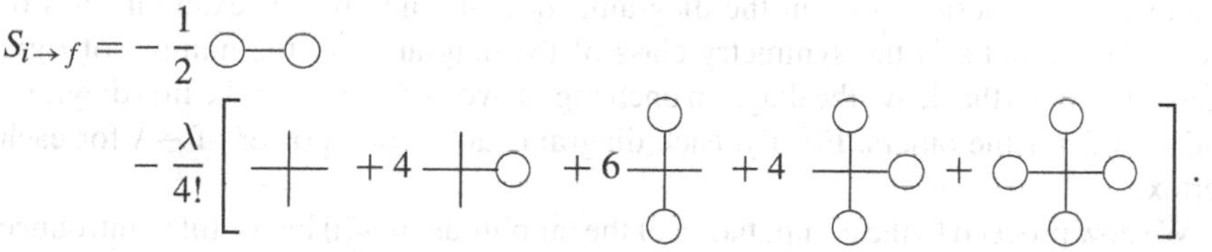

$$(4.2)$$

The crossing points of the lines in each diagram of this expansion are called *internal
vertices*; it is understood that each internal vertex corresponds to a space-time integration. Lines connecting two vertices are called *internal lines*, and correspond to a
factor of Δ (usually called the *propagator*) in the amplitude. The remaining lines are
external lines; each of them corresponds to a factor $\phi^{(as)}$.

C. M. Becchi and G. Ridolfi, *An Introduction to Relativistic Processes*
and the Standard Model of Electroweak Interactions, UNITEXT for Physics,
DOI: 10.1007/978-3-319-06130-6_4, © Springer International Publishing Switzerland 2014

The graphical form of Eq. (3.125) is

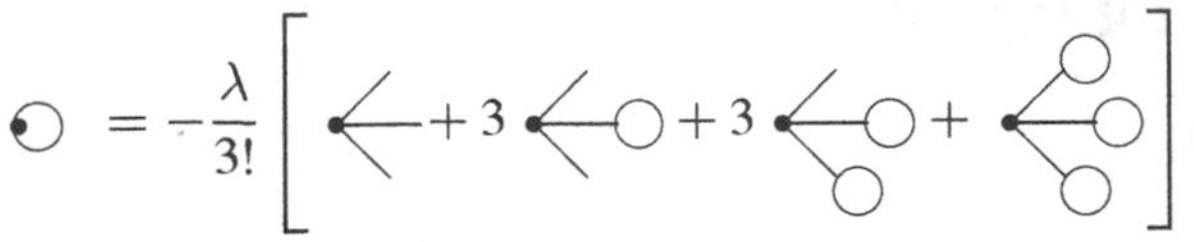

$$
\bigodot = -\frac{\lambda}{3!}\left[\ \prec + 3\ \prec\!\!\bigcirc + 3\ \prec\!\!\bigcirc + \ \prec\!\!\bigcirc\!\!\bigcirc\ \right], \tag{4.3}
$$

which can be solved iteratively:

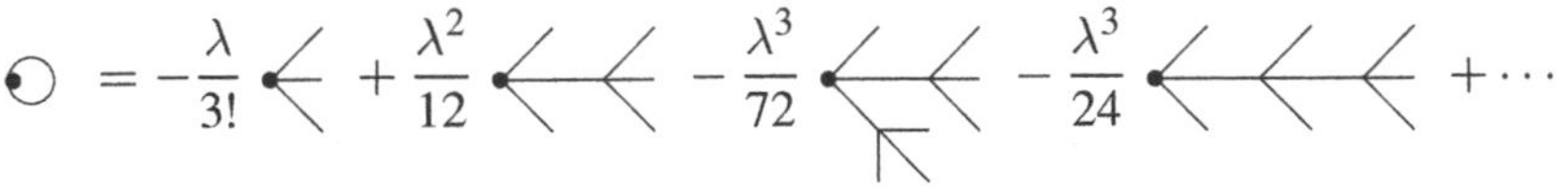

$$
\bigodot = -\frac{\lambda}{3!}\ \prec + \frac{\lambda^2}{12}\ \prec\!\!\prec - \frac{\lambda^3}{72}\ \prec\!\!\prec - \frac{\lambda^3}{24}\ \prec\!\!\prec\!\!\prec + \cdots . \tag{4.4}
$$

Using Eq. (4.4) in Eq. (4.2) we find

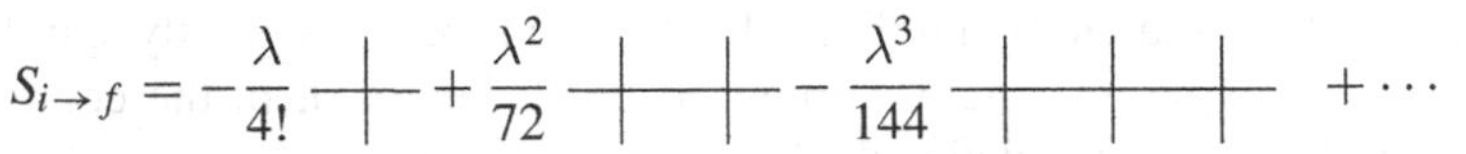

$$
S_{i\to f} = -\frac{\lambda}{4!}\ \vdash\!\!\vdash + \frac{\lambda^2}{72}\ \vdash\!\!\vdash\!\!\vdash - \frac{\lambda^3}{144}\ \vdash\!\!\vdash\!\!\vdash\!\!\vdash + \cdots \tag{4.5}
$$

The numerical coefficients in Eq. (4.5) are combinatorial factors, that can be written as

$$
\frac{1}{k\,\prod_v n_v!}, \tag{4.6}
$$

where v is a generic vertex in the diagram, n_v is the number of external lines of the vertex v, and k is the symmetry class of the diagram, i.e. the number of rigid transformations that leave the diagram unchanged: we find $k = 1$ for the first diagram, and $k = 2$ for the others. Finally, each diagram carries one power of $-\lambda$ for each vertex.

We now proceed to the computation of the amplitude. It will be useful to introduce four-dimensional Fourier transforms, defined as

$$
\hat{F}(k) \equiv \frac{1}{(2\pi)^4}\int d^4x\, e^{ik\cdot x}\, F(x) \qquad F(x) = \int d^4k\, e^{-ik\cdot x}\, \hat{F}(k) \tag{4.7}
$$

for a generic $F(x)$. We find

$$
\hat{\phi}(q) = \hat{\phi}^{(\mathrm{as})}(q) + \frac{1}{m^2 - q^2 - i\epsilon}\, \hat{J}(q), \tag{4.8}
$$

where, in the limit $\delta \to 0$ of infinite momentum resolution,

$$\hat{\phi}^{(\mathrm{as})}(q) = \frac{(\sqrt{4\pi}\delta)^{\frac{3}{2}}}{(2\pi)^{\frac{3}{2}}} \left[\sum_{i=1}^{2} \frac{\delta^{(4)}(q - p_i)}{\sqrt{2E_{\boldsymbol{p}_i}}} + \sum_{j=1}^{n} \frac{\delta^{(4)}(q + k_j)}{\sqrt{2E_{\boldsymbol{k}_j}}} \right]. \tag{4.9}$$

Therefore

$$\begin{aligned} S_{i\to f} =& -\frac{1}{2} \int d^4k\, d^4k'\, (2\pi)^4\, \delta^{(4)}\left(k + k'\right) \hat{J}(k) \frac{1}{m^2 - k^2 - i\epsilon} \hat{J}(k') \\ & - \frac{\lambda}{4!} \int \prod_{l=1}^{4} d^4q_l \left[\hat{\phi}^{(\mathrm{as})}(q_l) + \frac{1}{m^2 - q_l^2 - i\epsilon} \hat{J}(q_l) \right] (2\pi)^4\, \delta^{(4)} \left(\sum_{l=1}^{4} q_l \right). \end{aligned} \tag{4.10}$$

Similarly, Eq. (3.125) becomes

$$\begin{aligned} \hat{J}(k) &= -\frac{\lambda}{3!} \int \frac{d^4x}{(2\pi)^4} e^{ik\cdot x} \prod_{l=1}^{3} \int d^4q_l\, e^{-iq_l\cdot x} \left[\hat{\phi}^{(\mathrm{as})}(q_l) + \frac{\hat{J}(q_l)}{m^2 - q_l^2 - i\epsilon} \right] \\ &= -\frac{\lambda}{3!} \int \prod_{l=1}^{3} d^4q_l \left[\hat{\phi}^{(\mathrm{as})}(q_l) + \frac{\hat{J}(q_l)}{m^2 - q_l^2 - i\epsilon} \right] \delta^{(4)} \left(k - \sum_{l=1}^{3} q_l \right). \end{aligned} \tag{4.11}$$

The iterative expression of Eq. (4.5) in terms of Fourier transforms of the various quantities then reads

$$\begin{aligned} S_{i\to f} =& -\frac{\lambda}{4!} \int \left[\prod_{l=1}^{4} d^4q_l\, \hat{\phi}^{(\mathrm{as})}(q_l) \right] (2\pi)^4\, \delta^{(4)} \left(\sum_{l=1}^{4} q_l \right) \\ & + \frac{\lambda^2}{72} \int \left[\prod_{l=1}^{6} d^4q_l\, \hat{\phi}^{(\mathrm{as})}(q_l) \right] (2\pi)^4\, \delta^{(4)} \left(\sum_{l=1}^{6} q_l \right) \frac{1}{m^2 - \left(\sum_{l=1}^{3} q_l \right)^2 - i\epsilon} \\ & + \cdots \end{aligned} \tag{4.12}$$

By comparing Eq. (4.12) with its graphical form, Eq. (4.5), and taking Eq. (4.9) into account, we note that each line in the diagrams corresponds to a four-momentum integration variable; these momenta flow from initial to final external lines, and are constrained by momentum conservation at each vertex. The momenta carried by external lines are identified with the momenta of asymptotic initial (p_i) and final (k_j) state particles, while momentum conservation fixes the momenta of internal lines. The contribution of a given diagram to the amplitude is therefore the product of factors

$$\frac{1}{m^2 - q_l^2 - i\epsilon} \tag{4.13}$$

corresponding to internal lines, of a factor $-\lambda$ for each vertex, and a factor

$$\frac{(\sqrt{4\pi}\delta)^{\frac{3}{2}}}{\sqrt{(2\pi)^3\, 2E}} \tag{4.14}$$

for each external line. Finally, the amplitude is proportional to the factor

$$(2\pi)^4\, \delta^{(4)}\left(\sum_{j=1}^{n} k_j - \sum_{i=1}^{2} p_i\right) \tag{4.15}$$

that appears explicitly in Eq. (4.12). Note that the δ dependence of $S_{i\to f}$ in Eq. (3.75) is recovered if the momentum conservation delta function $\delta^{(4)}(P)$ is replaced by $\frac{e^{-P^2/\delta^2}}{\pi^2\delta^4}$. Summing over all possible assignments of asymptotic particles to external lines, one can check that every diagram appears with a multiplicity which is given by the number of permutations of the n_v asymptotic particles entering a given vertex v, multiplied by the number k of symmetry transformations of the diagram into itself. This gives a factor $k \prod_v n_v!$, that compensates exactly the denominator that appears explicitly in each term of Eq. (4.5).

Let us consider, as an example, the case of two particles in the final state, usually called elastic scattering:

$$\phi(p_1) + \phi(p_2) \to \phi(k_1) + \phi(k_2). \tag{4.16}$$

The relevant term is the first one in Eq. (4.12). We get

$$\begin{aligned} \mathcal{A}^{22}_{\boldsymbol{kp}} &= -i\frac{\lambda}{4!}\int\left[\prod_{l=1}^{4} d^4q_l\, \hat{\phi}^{(\mathrm{as})}(q_l)\right](2\pi)^4\,\delta^{(4)}\left(\sum_{l=1}^{4} q_l\right) \\ &= -i\,\frac{\lambda}{4!}\frac{4!(2\pi)^4\delta^{(4)}\,(p_1+p_2-k_1-k_2)}{\sqrt{\prod_{i=1}^{2}\left[(2\pi)^3 2E_{\boldsymbol{p}_i}\right]\prod_{j=1}^{2}\left[(2\pi)^3 2E_{\boldsymbol{k}_j}\right]}}\,(\sqrt{4\pi}\delta)^6. \end{aligned} \tag{4.17}$$

The factor 4! accounts for the number of ways one can extract terms linear in each plane wave from $\hat{\phi}^{(\mathrm{as})}(q_1)\ldots\hat{\phi}^{(\mathrm{as})}(q_4)$.

The function T, defined in Eq. (3.26), which appears in the general expression for the cross section, is now easily identified: since from Eq. (3.29)

$$\mathcal{A}^{22}_{\boldsymbol{kp}} = -2\pi i\, T(\boldsymbol{k}_1, \boldsymbol{k}_2; \boldsymbol{p}_1, \boldsymbol{p}_2)\,\delta^{(4)}(p_1+p_2-k_1-k_2)(\sqrt{4\pi}\delta)^6, \tag{4.18}$$

comparing with Eq. (4.17) we find

$$T(\boldsymbol{k}_1, \boldsymbol{k}_2; \boldsymbol{p}_1, \boldsymbol{p}_2) = \frac{\lambda}{\sqrt{4E_{\boldsymbol{p}_1}E_{\boldsymbol{p}_2}}}\frac{1}{\sqrt{(2\pi)^3 2E_{\boldsymbol{k}_1}}\sqrt{(2\pi)^3 2E_{\boldsymbol{k}_2}}}. \tag{4.19}$$

The differential cross section in given by

$$\begin{aligned} d\sigma_2 &= \frac{|\lambda|^2}{4E_{\boldsymbol{p}_1}E_{\boldsymbol{p}_2}}\frac{1}{|\boldsymbol{v}_1 - \boldsymbol{v}_2|}(2\pi)^4\,\delta^{(4)}(k_1 + k_2 - p_1 - p_2)\frac{d^3k_1}{(2\pi)^3 2E_{\boldsymbol{k}_1}}\frac{d^3k_2}{(2\pi)^3 2E_{\boldsymbol{k}_2}} \\ &= \frac{|\lambda|^2}{4\left|\boldsymbol{p}_1 E_{\boldsymbol{p}_2} - \boldsymbol{p}_2 E_{\boldsymbol{p}_1}\right|}(2\pi)^4\,\delta^{(4)}(k_1 + k_2 - p_1 - p_2)\frac{d^3k_1}{(2\pi)^3 2E_{\boldsymbol{k}_1}}\frac{d^3k_2}{(2\pi)^3 2E_{\boldsymbol{k}_2}}. \end{aligned} \tag{4.20}$$

The integration measure

$$d\phi_2 = (2\pi)^4\,\delta^{(4)}(k_1 + k_2 - p_1 - p_2)\frac{d^3k_1}{(2\pi)^3 2E_{\boldsymbol{k}_1}}\frac{d^3k_2}{(2\pi)^3 2E_{\boldsymbol{k}_2}} \tag{4.21}$$

is usually called the *invariant phase space*; it appears in the expression of the differential cross section for any process with two particles in the final state. Thanks to its transformation properties under Lorentz transformations, the invariant phase space can be computed in any reference frame. In many cases, a convenient choice is the rest frame of the center of mass of the particles in the initial state, where $\boldsymbol{p}_1 + \boldsymbol{p}_2 = 0$. In this frame,

$$\begin{aligned} d\phi_2 &= \frac{1}{(2\pi)^2}\,\delta^{(3)}(\boldsymbol{k}_1 + \boldsymbol{k}_2)\,\delta(\sqrt{s} - E_{\boldsymbol{k}_1} - E_{\boldsymbol{k}_2})\frac{d^3k_1\,d^3k_2}{2E_{\boldsymbol{k}_1}\,2E_{\boldsymbol{k}_2}} \\ &= \frac{1}{(2\pi)^2}\,\delta(\sqrt{s} - E_{\boldsymbol{k}_1} - E_{\boldsymbol{k}_2})\frac{d^3k_1}{4E_{\boldsymbol{k}_1}E_{\boldsymbol{k}_2}}, \end{aligned} \tag{4.22}$$

where we have defined $s = (p_1 + p_2)^2 = (E_{\boldsymbol{p}_1} + E_{\boldsymbol{p}_2})^2$. In the second step we have used the spatial momentum conservation factor $\delta^{(3)}(\boldsymbol{k}_1 + \boldsymbol{k}_2)$ to perform the integral over $\boldsymbol{k}_2$, therefore implicitly setting $\boldsymbol{k}_2 = -\boldsymbol{k}_1$ in the integrand. We may further simplify this expression using polar coordinates $\boldsymbol{k}_1 = (|\boldsymbol{k}_1|, \theta, \phi)$ and choosing the z axis in the direction of the momenta of incoming particles. Because of the symmetry of the whole process under rotations around the z axis, the squared amplitude cannot depend on the azimuthal angle ϕ. Hence,

$$d\phi_2 = \frac{1}{2\pi}\,\delta(\sqrt{s} - E_{\boldsymbol{k}_1} - E_{\boldsymbol{k}_2})\frac{|\boldsymbol{k}_1|^2\,d\,|\boldsymbol{k}_1|}{4E_{\boldsymbol{k}_1}E_{\boldsymbol{k}_2}}\,d\cos\theta. \tag{4.23}$$

The integral over $|\boldsymbol{k}_1|$ can now be performed using the delta function

$$\delta(\sqrt{s} - E_{k_1} - E_{k_2}) = \frac{E_{k_1} E_{k_2}}{\sqrt{s}\,|\boldsymbol{k}_1|}\,\delta\left(|\boldsymbol{k}_1| - \frac{\sqrt{s}}{2}\sqrt{1 - \frac{4m^2}{s}}\right). \tag{4.24}$$

We get

$$d\phi_2 = \frac{\beta}{16\pi}\, d\cos\theta, \tag{4.25}$$

where

$$\beta = \frac{|\boldsymbol{k}_1|}{E_{k_1}} = \sqrt{1 - \frac{4m^2}{s}} \tag{4.26}$$

is the velocity of the outgoing particle in the center-of-mass frame. An explicitly invariant form of the phase space measure can be obtained by defining the usual Mandelstam variable

$$t = (p_1 - k_1)^2. \tag{4.27}$$

In the center-of-mass frame we have

$$t = 2m^2 - \frac{s}{2}\left(1 - \beta^2 \cos\theta\right); \qquad dt = \frac{s}{2}\beta^2 d\cos\theta \tag{4.28}$$

and therefore

$$d\phi_2 = \frac{1}{8\pi}\,\frac{1}{\beta}\,\frac{dt}{s}. \tag{4.29}$$

The differential cross section is therefore given by

$$d\sigma_2 = \frac{\lambda^2}{32\pi s}\sqrt{1 - \frac{4m^2}{s}}\, d\cos\theta. \tag{4.30}$$

The total cross section is obtained by integrating $d\sigma_2$ in $\cos\theta$, recalling that in this case only one half of the total solid angle contributes, because of the identity of the particles in the final state.

4.2 The Invariant Amplitude

The calculations of the previous section can be immediately generalized to the case of a generic process with n particles in the final state. It is convenient to define the *invariant amplitude* $\mathcal{M}_{fi}$ through

$$\mathcal{A}^{n2}_{\boldsymbol{kp}} = i\,(2\pi)^4\,\delta^{(4)}\left(\sum_{j=1}^{n} k_j - \sum_{i=1}^{2} p_i\right) \frac{\mathcal{M}_{fi}(\sqrt{4\pi}\delta)^{\frac{3}{2}(n+2)}}{\sqrt{\prod_{i=1}^{2}\left[(2\pi)^3 2E_{\boldsymbol{p}_i}\right]\prod_{j=1}^{n}\left[(2\pi)^3 2E_{\boldsymbol{k}_j}\right]}}. \tag{4.31}$$

In the case of elastic scattering, the explicit calculation gives $\mathcal{M}_{2\to 2} = -\lambda$.

The function $T(\boldsymbol{k}, \boldsymbol{p})$ is related to the invariant amplitude by comparison with its definition:

$$\mathcal{A}^{n2}_{\boldsymbol{kp}} = -2\pi i\, T(\boldsymbol{k}_1, \dots, \boldsymbol{k}_n;\, \boldsymbol{p}_1, \boldsymbol{p}_2)\,(\sqrt{4\pi}\delta)^{\frac{3}{2}(n+2)}\delta^{(4)}(p_1 + p_2 - k_1 - \cdots - k_n). \tag{4.32}$$

We find

$$T(\boldsymbol{k}_1, \dots, \boldsymbol{k}_n;\, \boldsymbol{p}_1, \boldsymbol{p}_2) = -\frac{\mathcal{M}_{fi}}{\sqrt{4E_{\boldsymbol{p}_1}E_{\boldsymbol{p}_2}}}\frac{1}{\prod_{j=1}^{n}\sqrt{(2\pi)^3 2E_{\boldsymbol{k}_j}}}. \tag{4.33}$$

The differential cross section is therefore given by

$$d\sigma_n = \frac{\left|\mathcal{M}_{fi}\right|^2}{4\left|\boldsymbol{p}_1 E_{\boldsymbol{p}_2} - \boldsymbol{p}_2 E_{\boldsymbol{p}_1}\right|}\, d\phi_n(p_1, p_2; k_j), \tag{4.34}$$

where

$$d\phi_n(p_1, p_2; k_j) = (2\pi)^4\,\delta^{(4)}\left(\sum_{j=1}^{n} k_j - \sum_{i=1}^{2} p_i\right)\prod_{j=1}^{n}\frac{d^3 k_j}{(2\pi)^3 2E_{\boldsymbol{k}_j}} \tag{4.35}$$

is the invariant phase space for n particles in the final state. Finally, we observe that in all reference frames in which $\boldsymbol{p}_1$ and $\boldsymbol{p}_2$ have the same direction, we have

$$\left|\boldsymbol{p}_1 E_{\boldsymbol{p}_2} - \boldsymbol{p}_2 E_{\boldsymbol{p}_1}\right| = \sqrt{(p_1 \cdot p_2)^2 - m_1^2 m_2^2} \quad \text{when } \boldsymbol{p}_1 \parallel \boldsymbol{p}_2, \tag{4.36}$$

where m_1 and m_2 are the masses of initial state particles.

The same reduction to an invariant amplitude can be performed in the case of particle decays. Notice, first of all, that the matrix element Eq. (3.63) corresponds, up to a factor $-2\pi i\delta(E_f - E_i)$, to the linear part in H_W of the transition amplitude from a single-particle initial state to an n−particle final state. This amplitude can be computed in the semi-classical approximation using the method of Feynman diagrams.

Therefore, the general result has the form

$$T(\boldsymbol{k}_1, \dots, \boldsymbol{k}_n;\, \boldsymbol{p}) = -\sqrt{(2\pi)^3}\,\frac{\mathcal{M}_{fi}}{\sqrt{2E_{\boldsymbol{p}}}}\frac{1}{\prod_{j=1}^{n}\sqrt{(2\pi)^3 2E_{\boldsymbol{k}_j}}}. \tag{4.37}$$

where $\mathcal{M}_{fi}$ corresponds to the sum of Feynman diagrams with n particles in the final state and a single heavy particle in the initial state, with external line factors omitted. Inserting Eq. (4.37) into Eq. (3.65) yields

$$d\Gamma = \frac{|\mathcal{M}_{fi}|^2}{2E_p} d\phi_n(p; k_1, \ldots, k_n). \tag{4.38}$$

Needless to say, E_p equals the mass of the decaying particle in its rest frame, which is usually the most convenient choice for the computation of decay rates.

4.3 Feynman Rules for the Scalar Theory

The results obtained so far can be summarized as follows. In the context of the scalar theory defined by the interaction Lagrangian

$$\mathcal{L}_I = -\frac{\lambda}{4!}\phi^4 \tag{4.39}$$

the invariant amplitude $\mathcal{M}_{fi}$ for a generic process with two particles in the initial state and $2n$ particles in the final state is obtained, in the semi-classical approximation, by

1. selecting all connected diagrams with $2n + 2$ external lines and no closed loops that can be built with four-line vertices (note that $\mathcal{L}_I$ in Eq. (4.39) is proportional to the fourth power of ϕ);
2. assigning the asymptotic particles to the external lines of each diagram in all possible ways, and determining the momentum flux by means of four-momentum conservation at each vertex;
3. summing over all diagrams and over all possible assignments the product of $-\lambda$ to the power n_v, which is the number of vertices (equal to n in the present case), times a factor $\frac{1}{m^2 - q_l^2 - i\epsilon}$ for each internal line, where the four-momenta q_l are fixed by four-momentum conservation at each vertex.

We see that the essential ingredients of the calculation are propagators and vertex factors; these are characteristic of the particular theory one is considering. In particular, the propagators are the Green functions of the free field equations, while the vertex factors can be extracted from the interaction Lagrangian (Eq. (4.39) in our example) by computing the invariant amplitude for a virtual process in which the lines entering a given vertex are considered as external lines. It is easy to see that this general rule gives the correct result in the case of Eq. (4.39): each vertex has four lines, and the corresponding virtual amplitude is given by $-\lambda/4!$ times the number of possible assignments of external lines, 4!. This gives the correct factor of $-\lambda$. In the following, we will encounter more complicated cases, where the vertex factor may depend on particle momenta.

In order to illustrate the above procedure, let us compute the invariant amplitude for the process with two particles in the initial state and four particles in the final state ($n = 2$), that corresponds to the second term in Eq. (4.5).

Inequivalent assignments of asymptotic particles to external lines are usually expressed by oriented diagrams, with the external lines corresponding to the initial particles (labelled by a and b) incoming from below (or from the left), and final state particles (labelled by 1, 2, 3 and 4) outgoing above (or to the right). In the case at hand, we have two diagrams:

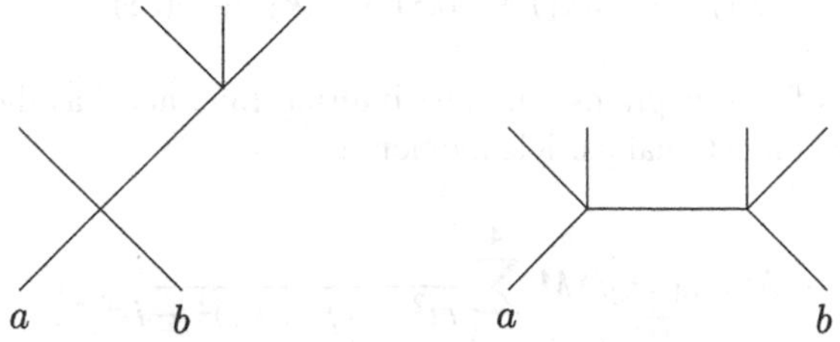

In the case of scalar particles, the orientation of internal lines is irrelevant. The diagram on the left corresponds to four inequivalent contributions, one for each choice $j = 1, \ldots, 4$ of the final state particle outgoing from the lower vertex; the momentum carried by the internal line is in this case $p_1+p_2-k_j$. The diagram on the right corresponds to six inequivalent contributions, that correspond to the six ways of choosing the two particles outgoing from the left vertex ($i < j = 2, 3, 4$); the momentum of the internal line is in this case $p_1 - k_i - k_j$. Summing all contributions according to the rules given above, we find

$$\mathcal{M}_{2\to4} = \lambda^2 \left[\sum_{j=1}^{4} \frac{1}{m^2 - \left(p_1 + p_2 - k_j\right)^2} + \sum_{j=2}^{4} \sum_{i=1}^{j-1} \frac{1}{m^2 - \left(p_1 - k_i - k_j\right)^2} \right]. \tag{4.40}$$

Note that we have not included the $i\epsilon$ terms in the internal line propagators; this is because for this particular amplitude the denominators never vanish for physical values of external particle four-momenta. For example, it is easy to show that four-momentum conservation implies $\left(p_1 + p_2 - k_j\right)^2 > 9m^2$. In general, this is always the case for theories that do not involve unstable particles.

In order to provide a sample calculation of a differential decay rate, we should consider a theory that describes particles with different masses. To this purpose, we extend our simple scalar model to include a second scalar field Φ, associated with particles of mass $M \gg m$, coupled to ϕ through an interaction term of the form

$$\mathcal{L}_I = -\frac{gM}{2}\, \Phi\, \phi^2, \tag{4.41}$$

where g is a new coupling constant (assumed small) and the factor of M was introduced to keep g dimensionless. To first order in g, the invariant amplitude for the two-body decay

$$\Phi(p) \to \phi(k_1) + \phi(k_2) \tag{4.42}$$

is simply given by

$$\mathcal{M}_{1\to 2} = -gM \tag{4.43}$$

while the invariant amplitude for

$$\Phi(p) \to \phi(k_1) + \phi(k_2) + \phi(k_3) + \phi(k_4) \tag{4.44}$$

gets contributions from diagrams with one internal line, and has therefore a non-trivial dependence on external particle momenta:

$$\mathcal{M}_{1\to 4} = g\lambda M \sum_{j=1}^{4} \frac{1}{m^2 - (p - k_j)^2 - i\epsilon}. \tag{4.45}$$

This concludes our construction of field theory scattering amplitudes in the semi-classical approximation. It is convenient at this point to state how our results are related to those of the general relativistic scattering theory. This starts from the assumption of some basic properties of the Hilbert space of asymptotic states, which is identified with a free-particle Fock space, and of the interacting field operators. Among these assumptions, locality plays a crucial role. Using these properties, Haag has shown that interacting field operators tend, for asymptotic times and in a suitable weak topology, to asymptotic free fields, which are built with the creation and annihilation operators of the particles in the asymptotic states. This result justifies the construction of relativistic scattering amplitudes in terms of Fourier transforms of suitable Green functions through the LSZ reduction formulae. In the most common approach, Green functions are identified with vacuum expectation values of time-ordered products of interacting fields. The collection of all Green functions can be cast into a functional generator that, in principle, can be computed using the Feynman functional integral formula. Our results are recovered by computing the Feynman functional integral in the saddle-point approximation. Here one finds again the result that characterizes the semi-classical approximation in path integral formulations of Quantum Mechanics. Radiative corrections correspond to corrections to the saddle point approximation.

4.4 Relativistic Particles in Matter

The presence of matter modifies the kinetic properties of particles, and in particular their mass, because of the coherent forward scattering of the relativistic particle by the matter. This effect has important consequences, e.g. in solar neutrino physics, and can be simply described in our framework.

Here we consider the simplified situation in which a relativistic particle, corresponding to the field ϕ, interacts with a uniform distribution of matter particles of mass M, corresponding to the field Φ, through the Lagrangian interaction term $-\frac{\lambda}{4}\phi^2\Phi^2$. We disregard any interaction among matter particles which occupy the same initial and final state corresponding to the uniform particle density ρ.

For the matter initial and final states we assume the simple form

$$|\Psi_N\rangle = \sqrt{\frac{1}{N!}}\frac{1}{\sqrt{(\sqrt{\pi}\delta)^3}}\left[\int d^3p e^{-\frac{p^2}{2\delta^2}} A^\dagger(\boldsymbol{p})\right]^N |\Omega\rangle, \tag{4.46}$$

where $A^\dagger(\boldsymbol{p})$ is a matter particle creation operator and $\rho = \frac{N\delta^3}{\pi^{\frac{3}{2}}}$, understanding the limit $N \to \infty$ and keeping ρ constant. This represents, in the limit, the mentioned uniform distribution of matter particles.

The kernel of the analysis is the study of the relativistic particle propagator in the presence of matter. As discussed in Sect. 3.4, the propagator, can be computed from the value of the action, $S_{i\to f}$, in the presence of an external source J_e coupled to the field ϕ. This source has been introduced in Eq. (3.90).

From the point of view of Feynman diagrams, the present theory is analogous to the $\lambda\phi^4$ theory considered so far. The only differences are that there are two different kinds of internal and external lines, and that the external lines corresponding to matter particles carry an extra factor $\sqrt{N}$ which follows immediately from the second quantization rules and from the form of the initial and final particle states.

In the present calculation, the relevant diagrams do not contain any ϕ external lines ($\phi^{(\mathrm{as})} = 0$), while the two J_e sources are coupled to the end points of ϕ internal lines. Indeed we are interested in the terms of $S_{i\to f}$ depending quadratically on J_e. Therefore, the interaction being quadratical in ϕ, the relevant diagrams contain a chain of ϕ lines joining two external sources J_e and there are no internal matter field lines. The internal vertices correspond to elementary interactions of the relativistic particle with matter; hence each vertex carries one incoming and one outgoing matter line.

Thus the amplitude $S_{i\to f}$ appears as the sum of the series of Feynman diagrams shown in Eq. (4.5) where, however, the vertical lines are matter field lines and the first and the last horizontal lines are replaced by ϕ internal lines coupled to J_e. Note that in this case the combinatorial factors are different from those of Eq. (4.6).

According to the Feynman rules, and in particular the external line factor Eq. (4.14), the contribution to the amplitude of the diagram with n vertices is

$$\begin{aligned}
&-\frac{(2\pi)^4}{2}\int d^4p\,\frac{\tilde{J}_e(-p)\tilde{J}_e(p)}{m^2-p^2-i\epsilon}\left[-\frac{(\sqrt{4\pi}\delta)^3N\lambda}{(2\pi)^3 2M}\frac{1}{m^2-p^2-i\epsilon}\right]^n \\
&= -\frac{(2\pi)^4}{2}\int d^4p\,\frac{\tilde{J}_e(-p)\tilde{J}_e(p)}{m^2-p^2-i\epsilon}\left[-\frac{\rho\lambda}{2M}\frac{1}{m^2-p^2-i\epsilon}\right]^n,
\end{aligned} \tag{4.47}$$

where the factors N appear as a consequence of the modified external line factor of matter particles. The series is easily summed, and we obtain the relativistic particle propagator

$$\Delta_{matter}(p) = \frac{1}{m^2 - p^2 - i\epsilon} \sum_{n=0}^{\infty} \left[-\frac{\rho\lambda}{2M} \frac{1}{m^2 - p^2 - i\epsilon} \right]^n$$
$$= \frac{1}{m^2 + \frac{\rho\lambda}{2M} - p^2 - i\epsilon}. \tag{4.48}$$

This simple calculation shows that the presence of matter introduces a correction to the squared mass of the relativistic particle proportional to the forward scattering amplitude of the relativistic particle by matter particles, times the matter particle density. This result turns out to be generally true.

4.5 Unitarity, Radiative Corrections and Renormalizability

The constraint in Eq. (3.22), that follows from unitarity of the scattering matrix, can be formulated in terms of invariant amplitudes, defined as in Eq. (4.33). We find

$$\mathcal{M}_{ij} - \mathcal{M}_{ji}^* = i \sum_f \int d\phi_{n_f}(P_i; k_1^f, \ldots, k_{n_f}^f)\, \mathcal{M}_{if} \mathcal{M}_{jf}^*, \tag{4.49}$$

with $d\phi_{n_f}$ given in Eq. (4.35). The above constraint is relevant to our analysis, because it is systematically violated in the semi-classical approximation, thus indicating that corrections to this approximation are needed. In the semi-classical approximation the amplitudes are real and symmetric under the exchange of initial and final states[1]; hence, the left-hand side of Eq. (4.49) vanishes, while the right-hand side does not. This can be seen explicitly in the simple case of two-particle elastic scattering in the context of the real scalar field theory. We have shown in Sect. 4.1 that the invariant amplitude in this case is, in the semiclassical approximation,

$$\mathcal{M}_{2\to 2}^{(0)} = -\lambda. \tag{4.50}$$

The left-hand side of the unitarity constraint Eq. (4.49) vanishes in the semi-classical approximation, since $\mathcal{M}_{2\to 2}^{(0)}$ is real. The right-hand side of Eq. (4.49) is easily computed if we further choose the center-of-mass energy to be smaller than $4m$, so that final states f with more than two scalars are kinematically forbidden. We get

[1] This is only true when the theory is time-reversal invariant.

$$i\lambda^2 \int d\phi_2(P_i; k_1, k_2) = \frac{i\lambda^2 \, |\boldsymbol{k}_1|}{16\pi E_{\boldsymbol{k}_1}} = \frac{i\lambda^2}{16\pi}\sqrt{1 - \frac{4m^2}{s}}, \tag{4.51}$$

where we have used Eq. (4.25) for the invariant two-particle phase space, and the integral has been restricted to one half of the total solid angle due to the identity of scalar particles. Equation (4.51) is manifestly nonzero, and the unitarity constraint is therefore violated.

Unitarity is recursively restored taking radiative corrections into account. In particular, the one-loop contribution corresponds to three different diagrams with two internal lines and two vertices, and hence one loop:

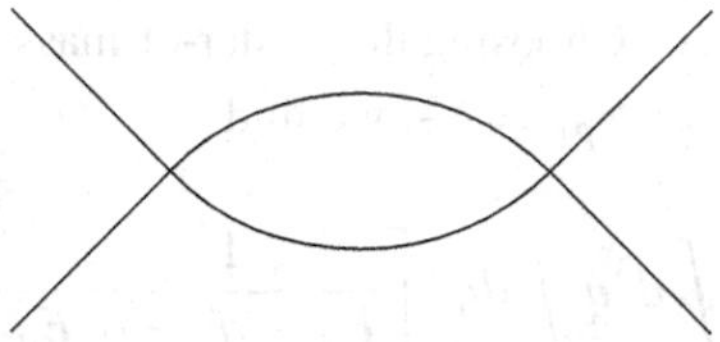

The three diagrams correspond to the three different possible assignments of external momenta to external lines.

The amplitudes corresponding to loop diagrams can be computed using the rules given in the previous section, with two further instructions: first, the integral over the loop momentum q, which is not fixed by momentum conservation at the vertices, must be performed with the measure

$$-i\,\frac{d^4q}{(2\pi)^4}, \tag{4.52}$$

and second, the amplitude carries a combinatorial factor, given by the inverse of the number of symmetry transformations of the diagram. In the present case this factor is $1/2$, because the diagram is symmetric under permutation of the internal lines.

The one-loop contribution to the $2 \to 2$ invariant amplitude is therefore

$$\mathcal{M}^{(1)}_{2\to 2} = \mathcal{M}(s) + \mathcal{M}(t) + \mathcal{M}(u), \tag{4.53}$$

where s, t, u are the usual Mandelstam invariants

$$s = (p_1 + p_2)^2; \qquad t = (p_1 - k_1)^2; \qquad u = (p_1 - k_2)^2, \tag{4.54}$$

and

$$\mathcal{M}(p^2) = -i\frac{\lambda^2}{2}\int \frac{d^4q}{(2\pi)^4}\left[\frac{1}{m^2 - q^2 - i\epsilon}\,\frac{1}{m^2 - (p-q)^2 - i\epsilon} - \frac{1}{M^2 - q^2 - i\epsilon}\,\frac{1}{M^2 - (p-q)^2 - i\epsilon}\right] - C(M^2). \tag{4.55}$$

Note that we have introduced in the integrand of Eq. (4.55) a term that depends on a large mass parameter M in order to regularize the loop integral, otherwise divergent in the large-momentum region. This procedure is called *Pauli-Villars regularization*; the parameter M is eventually taken to infinity. For any finite value of M, the integral of the difference of the two terms in Eq. (4.55) is convergent, while that of each single term diverges. The introduction of the regularizing term must be compensated by the real constants $C(M^2)$, whose contributions to the scattering amplitude appears as a correction (counter-term) to λ. The constants $C(M^2)$ must be tuned in such a way that the invariant scattering amplitude is equal to $-\lambda$ for some fixed value of the external momenta (for example, at the scattering threshold.) This is called a *renormalization prescription*.

Let us first compute $\mathcal{M}(s)$. Choosing the center-of-mass frame of the particles in the initial state, where $E_{\boldsymbol{k}_1} = E_{\boldsymbol{p}_1} = \frac{\sqrt{s}}{2}$, we find

$$\begin{aligned}\mathcal{M}(s) = &-i\frac{\lambda^2}{2(2\pi)^4}\int d^3q\int dq^0\left[\frac{1}{E_{\boldsymbol{q}}-q^0-i\epsilon}\frac{1}{E_{\boldsymbol{q}}+q^0-i\epsilon}\right.\\ &\times\left.\frac{1}{E_{\boldsymbol{q}}+\sqrt{s}-q^0-i\epsilon}\frac{1}{E_{\boldsymbol{q}}-\sqrt{s}+q^0-i\epsilon}-\mathrm{PV}\right]-C(M^2),\end{aligned} \tag{4.56}$$

where $E_{\boldsymbol{q}} = \sqrt{m^2+\boldsymbol{q}^2}$ and PV stands for an analogous fraction with $E_{\boldsymbol{q}}$ replaced by $\sqrt{M^2+\boldsymbol{q}^2}$. The q^0 integration is readily performed by the method of residues. Switching to polar coordinates for $\boldsymbol{q}$ we get

$$\mathcal{M}(s) = \frac{\lambda^2}{4(2\pi)^2}\int_0^\infty |\boldsymbol{q}|^2 d|\boldsymbol{q}|\left[\frac{1}{E_{\boldsymbol{q}}\left(E_{\boldsymbol{q}}^2-\frac{s}{4}-i\epsilon\right)}-\mathrm{PV}\right]-C(M^2). \tag{4.57}$$

It is now convenient to change the integration variable to

$$\sigma = 4E_{\boldsymbol{q}}^2. \tag{4.58}$$

We obtain

$$\begin{aligned}\mathcal{M}(s) &= \frac{\lambda^2}{8(2\pi)^2}\left[\int_{4m^2}^{\infty} d\sigma\sqrt{1-\frac{4m^2}{\sigma}}\frac{1}{\sigma-s-i\epsilon}-\mathrm{PV}\right]-C(M^2)\\ &= \frac{\lambda^2}{8(2\pi)^2}\left[\int_{4m^2}^{4M^2} d\sigma\sqrt{1-\frac{4m^2}{\sigma}}\frac{1}{\sigma-s-i\epsilon}-K(s)\right]-C(M^2),\end{aligned} \tag{4.59}$$

where

$$K(s) = \int_{4M^2}^{\infty} \frac{d\sigma}{\sigma - s - i\epsilon} \left(\sqrt{1 - \frac{4m^2}{\sigma}} - \sqrt{1 - \frac{4M^2}{\sigma}} \right)$$
$$= \int_{1}^{\infty} \frac{dx}{x - \frac{s}{4M^2} - i\epsilon} \left(\sqrt{1 - \frac{m^2}{M^2}\frac{1}{x}} - \sqrt{1 - \frac{1}{x}} \right) \tag{4.60}$$

is a finite number, which becomes independent of s as $M^2 \to \infty$, and can therefore be absorbed in a redefinition of the counter-term $C(M^2)$.

The analytic properties of $\mathcal{M}(s)$ are now immediately read off Eq. (4.59): the regularized integral gives a function of s which is real on the negative real axis and analytic in the complex plane, with a cut on the positive real axis in the range $4m^2 \leq s \leq 4M^2$. The cut discontinuity of this function is purely imaginary and equal to

$$2i \operatorname{Im} \mathcal{M}(s) = \mathcal{M}(s) - \mathcal{M}^*(s) = \frac{i\lambda^2}{16\pi}\sqrt{1 - \frac{4m^2}{s}}. \tag{4.61}$$

This contribution to the left-hand side of Eq. (4.49) exactly equals Eq. (4.51). The remaining two terms in Eq. (4.53), $\mathcal{M}(t)$ and $\mathcal{M}(u)$, only contribute to the real part of the amplitude. This can be checked by rotating the q^0 integration path from the real to the imaginary axis of the complex plane, which is allowed since the rotating path does not cross any singularity of the integrand. Thus, we have seen that the semi-classical approximation violates $\mathcal{S}$-matrix unitarity, which is however recursively restored by taking into account the contributions of radiative corrections, i.e. of Feynman diagrams with closed loops.

To complete the calculation of the one-loop corrections to the $2 \to 2$ amplitude, we observe that

$$\mathcal{M}(s) = \frac{1}{\pi} \int_{4m^2}^{4M^2} d\sigma \, \frac{\operatorname{Im} \mathcal{M}(\sigma)}{\sigma - s - i\epsilon} - C(M^2), \tag{4.62}$$

which is a dispersion relation cut-off at $4M^2$.

The condition that $\mathcal{M}(s)$ be finite implies that the $C(M^2)$ behaves as

$$C(M^2) \simeq \frac{\lambda^2}{(4\pi)^2} \ln \frac{M}{m} \tag{4.63}$$

at large M. If we further require that the loop corrections vanish when $p_1 = p_2 = k_1 = k_2 = 0$ we get in the limit $M^2 \to \infty$

$$\mathcal{M}(s) = \frac{\lambda^2}{8(2\pi)^2} \int_{4m^2}^{\infty} d\sigma \sqrt{1 - \frac{4m^2}{\sigma}} \left[\frac{1}{\sigma - s - i\epsilon} - \frac{1}{\sigma} \right]$$

$$= \frac{s}{\pi} \int_{4m^2}^{\infty} \frac{d\sigma}{\sigma} \frac{\operatorname{Im} \mathcal{M}(\sigma)}{\sigma - s - i\epsilon} \tag{4.64}$$

which is a (once) subtracted dispersion relation, and defines the renormalized amplitude with the prescription that the coupling constant λ be equal to minus the invariant amplitude at zero external momenta. This prescription amounts to defining

$$C(M^2) = \frac{\lambda^2}{8(2\pi)^2} \int_{4m^2}^{4M^2} \frac{d\sigma}{\sigma} \sqrt{1 - \frac{4m^2}{\sigma}}. \tag{4.65}$$

The above example shows that covariance is a crucial ingredient of renormalization. Secondly, we have seen how the dispersion relations allow us to cut off the divergent integrals and, finally, how renormalized Feynman amplitudes correspond to subtracted dispersion relations. The approach to renormalization based on dispersion relations is very simple at the one-loop level; it is however rarely employed nowadays, because it becomes cumbersome when applied to more complex diagrams.

In conclusion, we have seen in the example that the four-momenta of the loop lines are not completely determined by the four-momentum conservation at the vertices and therefore the amplitudes contain integrations over undetermined momenta. Such integrals can be divergent in the region of large momenta (the so called *ultraviolet* region.) In this situation one must introduce a regularization procedure which must be accompanied by the introduction of counter-terms defined up to additive constants. In the case of elastic scattering this constant is reabsorbed into a redefinition of the coupling λ, but in general this leads to the appearance of new physical parameters. If the process does not come to an end, the theory fails to be predictive, since the number of input data diverges. When this happens the theory is called non-renormalizable.

There is a close connection between the properties of the interaction terms in the Lagrangian density and the strength of divergences of loop integrals. This can be seen using pure dimensional analysis. In natural units, there is only one independent scale, say the scale of energies or masses. Since the Lagrangian density scales as the fourth power of mass, bosonic fields have the dimensionality of a mass, so that $(\partial\phi)^2$ has the correct dimensionality. For the same reason, an interaction term of degree D in the fields and field derivatives is proportional to a coefficient with the dimension of a mass to the power $4 - D$. Thus, for example, an interaction term proportional to ϕ^6, or else $\phi^2(\partial\phi)^2$, appears in the Lagrangian density with a coefficient of dimension (mass)$^{-2}$. For a given amplitude, diagrams with loops must have the same dimensionality as diagrams in the semi-classical approximation; the presence of coefficients with the dimension of negative powers of an energy must be compensated by positive powers of integration four-momenta, therefore making the ultraviolet behaviour of the integral worse. It is possible to show that a field theory whose Lagrangian density is a generic linear combination of all possible monomials in the fields and their derivatives of dimensionality $D \leq 4$ is renormalizable in the sense that the problem of divergent integrals in the ultraviolet region can be solved by suitable redefinitions of the parameters in the action. This result is sufficient to guarantee the physical predictivity of field theories involving scalars fields, and it is immediately extended to fermions. When vector fields are present, further difficulties arise, as we shall see in Chap. 8.

Chapter 5
Spinor Fields

5.1 Spinor Representations of the Lorentz Group

We have discussed in Sect. 2.1 the importance of relativistic invariance in the formulation of theories of fundamental interactions. We have studied in detail the case of scalar fields, transforming as

$$\phi'(x) = \phi(\Lambda^{-1}x) \tag{5.1}$$

under a Lorentz transformation Λ. A second well-known example (which will be discussed at length in Chaps. 7 and 8) is provided by vector fields:

$$A'^{\mu}(x) = \Lambda^{\mu}_{\nu}\, A^{\nu}(\Lambda^{-1}x). \tag{5.2}$$

We now consider the general case, when a system of complex fields $\phi_\alpha,\ \alpha = 1, \ldots, n$ transform as

$$\phi'_\alpha(x) = \sum_{\beta=1}^{n} S(\Lambda)_\alpha^\beta \phi_\beta(\Lambda^{-1}x), \tag{5.3}$$

under a Lorentz transformation. The matrix $S(\Lambda)$ obeys the condition

$$S(\Lambda\Lambda') = S(\Lambda)S(\Lambda') \tag{5.4}$$

for any two Lorentz transformations Λ, Λ'. The vector field is an obvious example; in that case, we have simply $S(\Lambda) = \Lambda$.

In this section, we shall discuss a less trivial example of a realization of Eq. (5.3). We observe that 2×2 hermitian matrices are in one-to-one correspondence with four vectors. Indeed, a generic 2×2 hermitian matrix x can be parametrized as

$$x = \begin{pmatrix} x^0 + x^3 & x^1 - ix^2 \\ x^1 + ix^2 & x^0 - x^3 \end{pmatrix} \equiv x^\mu \sigma_\mu \tag{5.5}$$

C. M. Becchi and G. Ridolfi, *An Introduction to Relativistic Processes and the Standard Model of Electroweak Interactions*, UNITEXT for Physics,
DOI: 10.1007/978-3-319-06130-6_5,

where

$$\sigma_0 = \begin{pmatrix} 1 & 0 \\ 0 & 1 \end{pmatrix} \quad \sigma_1 = \begin{pmatrix} 0 & 1 \\ 1 & 0 \end{pmatrix} \quad \sigma_2 = \begin{pmatrix} 0 & -i \\ i & 0 \end{pmatrix} \quad \sigma_3 = \begin{pmatrix} 1 & 0 \\ 0 & -1 \end{pmatrix}. \tag{5.6}$$

We have

$$\det x = \left(x^0\right)^2 - \left(x^1\right)^2 - \left(x^2\right)^2 - \left(x^3\right)^2 \equiv x^\mu g_{\mu\nu} x^\nu = x^\mu x_\mu. \tag{5.7}$$

Let us consider a generic 2×2 complex matrix L with unit determinant. These matrices form a group, which is usually called $SL(2C)$:

$$L = \begin{pmatrix} a & b \\ c & d \end{pmatrix}; \quad \det L = ad - bc = 1. \tag{5.8}$$

Such a matrix is completely specified by six real parameters, the same number as the parameters in Λ. The linear transformation

$$x \to x' = LxL^\dagger \tag{5.9}$$

leaves both $\det x$ and the sign of x^0 unchanged. Furthermore, x' is also hermitian. Thus, this transformation defines a Lorentz transformation

$$x^\mu \to x'^\mu = \Lambda^\mu_\nu(L)\, x^\nu. \tag{5.10}$$

In order to compute the matrix Λ that corresponds to a given L, we observe that

$$LxL^\dagger = L\sigma_\nu L^\dagger x^\nu = \sigma_\mu x'^\mu = \sigma_\mu \Lambda^\mu_\nu(L) x^\nu \tag{5.11}$$

or

$$L\sigma_\nu L^\dagger = \sigma_\mu \Lambda^\mu_\nu. \tag{5.12}$$

It will be convenient to introduce matrices $\bar\sigma_\mu$ defined as

$$\bar\sigma_0 = \sigma_0; \qquad \bar\sigma_i = -\sigma_i. \tag{5.13}$$

It is easy to check that

$$\mathrm{Tr}\left(\bar\sigma_\mu \sigma_\nu\right) = 2g_{\mu\nu}. \tag{5.14}$$

Therefore

$$\frac{1}{2}\mathrm{Tr}\left(\bar{\sigma}_\mu L \sigma_\nu L^\dagger\right) = g_{\mu\rho}\Lambda^\rho_\nu, \tag{5.15}$$

or equivalently

$$\frac{1}{2}\mathrm{Tr}\left(\bar{\sigma}^\mu L \sigma_\nu L^\dagger\right) = \Lambda^\mu_\nu. \tag{5.16}$$

Consider as an example the matrix

$$L = \begin{pmatrix} e^{\frac{\chi - i\theta}{2}} & 0 \\ 0 & e^{-\frac{\chi - i\theta}{2}} \end{pmatrix}. \tag{5.17}$$

An explicit calculation shows that the corresponding $\Lambda(L)$ is given by

$$\Lambda(L) = \begin{pmatrix} \cosh\chi & 0 & 0 & \sinh\chi \\ 0 & \cos\theta & -\sin\theta & 0 \\ 0 & \sin\theta & \cos\theta & 0 \\ \sinh\chi & 0 & 0 & \cosh\chi \end{pmatrix}, \tag{5.18}$$

which represents a rotation by an angle θ about the z axis, and a boost with velocity $\beta = \tanh\chi$ along the z axis.

The reader has certainly recognized in Eq. (5.17) with $\chi = 0$ the matrix that represents a rotation by an angle θ around the z axis for an ordinary spinor representing the quantum states of a spin-1/2 particle in ordinary quantum mechanics.

In a natural way, one would invert the relation Eq. (5.16) between Λ and L, and introduce a new kind of fields, called the *spinor* fields, transforming as

$$\xi'_\alpha(x) = \sum_{\beta=1}^{2} L(\Lambda)^\beta_\alpha\, \xi_\beta(\Lambda^{-1}x). \tag{5.19}$$

The relation Eq. (5.16), however, is not invertible, since

$$\Lambda(L) = \Lambda(-L), \tag{5.20}$$

and Eq. (5.19) is ill defined. This is a manifestation of the fact that spin rotation matrices are two-valued functions of the rotation angle.

A deeper understanding of this phenomenon arises from the observation that the rotation group (and therefore the Lorentz group) is a continuous group: the action of a rotation on a physical system or reference frame is a continuous operation, resulting from a sequence of infinitesimal transformations, that is, on the transformation path.

Classically, the effect of a rotation on a system is described by the relations between positions and velocities of points before and after the rotation. This clearly does not depend on the rotation path. On the other hand, in quantum mechanics one would expect the wave function of a generic quantum system to acquire a phase factor as a result of a rotation. The question is, whether and how this phase factor depends on the path.

Let us consider two different rotation paths with the same end-points. Let us assume that the wave function transforms as

$$\psi \to U_1\psi;\ \psi \to U_2\psi, \tag{5.21}$$

where U_1, U_2 are unitary transformations, corresponding to paths 1 or 2 respectively. Because the end-points coincide, the final states can at most differ by a phase factor:

$$U_2 = e^{i\phi}U_1, \tag{5.22}$$

Two such paths can be combined to build a closed path (which classically corresponds to a trivial rotation) by simply combining one of the two, say the second one, with the first one followed in the opposite direction. Then the closed rotation path transforms the wave function as

$$\psi \to U_1^\dagger U_2\psi = e^{i\phi}\psi. \tag{5.23}$$

This suggests that there might be an infinity of possible values of the phase ϕ, one for each closed path. This is however not the case; indeed, two different rotation paths with the same end-points which can be continuously deformed into one another must have the same action on wave functions.[1] For example, in the case of closed rotation paths which are topologically trivial, in the sense that they can be continuously reduced to a point, the phase ϕ must vanish.

The question which remains to be answered is whether there are any topologically non-trivial rotation paths. The answer to this question is in the affirmative: indeed, a 2π rotation around any axis is a closed rotation path, since the initial and final physical states coincide, but is topologically non-trivial: it can be continuously deformed into a 2π rotation around any other axis, but it cannot be continuously reduced to the trivial action.

On the contrary, a 4π rotation around any axis is topologically trivial, because it can be seen as two subsequent 2π rotations around the same axis, and the second 2π rotation can be continuously deformed into a 2π rotation around the opposite axis. Therefore the 4π rotation is topologically equivalent to a 2π back-and-forth rotation. Now it is apparent that a back-and-forth rotation of any angle θ corresponds to a topologically trivial path, since the whole path can be countinuously contracted to a point by reducing θ to zero.

We must conclude that, after a 2π rotation, a wave function gets a multiplicative phase factor $e^{i\phi}$. After a 4π rotation the phase becomes 2ϕ, which must be a multiple

[1] This is because the quantum action of a rotation must correspond to a unitary transformation with unit determinant. This excludes the possibility of continuous variations of the phase ϕ. Since the unitary transformation group of rotation paths is a continuous manifold, the phase must remain constant when the rotation path is continuously deformed.

of 2π because a 4π rotation is a topologically trivial path. Hence $\phi = n\pi$, or $e^{i\phi} = \pm 1$. In the case of scalar and vector fields $e^{i\phi(2\pi)} = 1$, but in the case of spinors one has $e^{i\phi(2\pi)} = -1$, and hence $L(\Lambda)$ is a two-valued function.

Therefore, the space-time transformation symmetry does not correspond to the Lorentz group, but rather to the group of the L matrices, $SL(2C)$. Finally, we note that the above considerations do not affect the theory of integer-spin fields, since in those cases the action of the Lorentz group and of $SL(2C)$ coincide.

Let us therefore introduce a complex two-component spinor field $\xi_R(x)$ that under a change of reference frame transforms as

$$\xi_R(x) \to \xi'_R(x) = L\,\xi_R\left(\Lambda^{-1}(L)x\right), \tag{5.24}$$

(the meaning of the suffix R will be clear soon), and let us build a Lorentz-invariant Lagrangian density out of the new fields and their derivatives. The Lagrangian density contains even powers of the spinor fields, otherwise it would change sign under a rotation by an angle 2π around any axis. We observe that

$$L^T\,\epsilon\,L = \epsilon, \tag{5.25}$$

where ϵ is the antisymmetric matrix

$$\epsilon = \begin{pmatrix} 0 & 1 \\ -1 & 0 \end{pmatrix}, \tag{5.26}$$

and T denotes the usual operation of matrix transposition. Indeed,

$$\begin{pmatrix} a & c \\ b & d \end{pmatrix}\begin{pmatrix} 0 & 1 \\ -1 & 0 \end{pmatrix}\begin{pmatrix} a & b \\ c & d \end{pmatrix} = (ad - bc)\begin{pmatrix} 0 & 1 \\ -1 & 0 \end{pmatrix} \tag{5.27}$$

and $ad - bc = \det L = 1$. Using Eqs. (5.12) and (5.25), one can prove that

$$\xi_R^\dagger\,\bar\sigma_\mu\,\xi_R \tag{5.28}$$

has the same transformation properties as a covariant four-vector:

$$\begin{aligned}
\xi_R^\dagger\,L^\dagger\,\bar\sigma_\mu\,L\xi_R &= -\xi_R^\dagger\,L^\dagger\,\epsilon\,\sigma_\mu^T\,\epsilon\,L\xi_R \\
&= -\xi_R^\dagger\,\epsilon\,(L^{-1})^*\,\sigma_\mu^T\,(L^{-1})^T\,\epsilon\,\xi_R \\
&= -\left(\Lambda^{-1}\right)_\mu^{\ \nu}\,\xi_R^\dagger\,\epsilon\,\sigma_\nu^T\,\epsilon\,\xi_R \\
&= \left(\Lambda^{-1}\right)_\mu^{\ \nu}\,\xi_R^\dagger\,\bar\sigma_\nu\,\xi_R.
\end{aligned} \tag{5.29}$$

where we have used

$$\epsilon\sigma_\mu\epsilon = -\bar\sigma_\mu^T. \tag{5.30}$$

Therefore, the bilinear form

$$\xi_R^\dagger \, \bar{\sigma}_\mu \, \partial^\mu \xi_R \tag{5.31}$$

transforms as a scalar under Lorentz transformations. This result allows us to build a Lagrangian density for ξ_R:

$$\mathcal{L}^{(R)}_{\text{Weyl}} = i \, \xi_R^\dagger \bar{\sigma}_\mu \, \partial^\mu \xi_R . \tag{5.32}$$

It is easy to check that $\mathcal{L}^{(R)}$ is hermitian, up to a total derivative, and that it transforms as a scalar field. The corresponding field equation for ξ_R is

$$\bar{\sigma}_\mu \partial^\mu \, \xi_R = 0, \tag{5.33}$$

This equation has negative-frequency plane-wave solutions

$$\xi_R^{(-)} = \frac{u(p)}{(2\pi)^{\frac{3}{2}}} e^{-i(|\boldsymbol{p}|t - \boldsymbol{p}\cdot\boldsymbol{r})}, \tag{5.34}$$

and positive-frequency solutions

$$\xi_R^{(+)} = \frac{v(p)}{(2\pi)^{\frac{3}{2}}} e^{i(|\boldsymbol{p}|t - \boldsymbol{p}\cdot\boldsymbol{r})}, \tag{5.35}$$

where $u(p)$ and $v(p)$ are such that

$$\boldsymbol{\sigma} \cdot \boldsymbol{p} \, u(\boldsymbol{p}) = |\boldsymbol{p}| \, u(\boldsymbol{p}) \tag{5.36}$$

$$\boldsymbol{\sigma} \cdot \boldsymbol{p} \, v(\boldsymbol{p}) = |\boldsymbol{p}| \, v(\boldsymbol{p}). \tag{5.37}$$

Using the identity

$$\epsilon \boldsymbol{\sigma}^T = \boldsymbol{\sigma} \epsilon \tag{5.38}$$

one can check that

$$u(\boldsymbol{p}) = \frac{1}{\sqrt{2|\boldsymbol{p}|}} \begin{pmatrix} \sqrt{|\boldsymbol{p}| + p^3} \\ \frac{p^1 + ip^2}{\sqrt{|\boldsymbol{p}| + p^3}} \end{pmatrix}; \qquad v(\boldsymbol{p}) = \epsilon u^*(-\boldsymbol{p}), \tag{5.39}$$

are solutions of Eqs. (5.36, 5.37), with the normalization

$$u^\dagger(\boldsymbol{p}) u(\boldsymbol{p}) = v^\dagger(\boldsymbol{p}) v(\boldsymbol{p}) = 1. \tag{5.40}$$

The free theory can be quantized in analogy with the case of the scalar field, with one important difference: the total energy in this case is found to be unbounded from below, because states with negative frequency correspond to negative energy.

The system would therefore be intrinsically unstable, unless one assumes that negative-energy states are all occupied and inaccessible because of the Pauli exclusion principle. For this reason, in the semi-classical limit spinor fields acquire an anti-commutativity property, so that any exchange operation on a given amplitude induces a sign flip. As a consequence, amplitudes are antisymmetric functions of momenta and spins of identical spinor particles.

Using the second-quantization formalism, we can express the above results by the following decomposition of the spinor field:

$$\begin{aligned}\xi_R(x) &= \frac{1}{(2\pi)^{\frac{3}{2}}}\int d^3p\left[u(\boldsymbol{p})\alpha_R(\boldsymbol{p})e^{-ip\cdot x} + v(\boldsymbol{p})\beta_R^\dagger(\boldsymbol{p})e^{ip\cdot x}\right] \\ &= \frac{1}{(2\pi)^{\frac{3}{2}}}\int d^3p\left[u(\boldsymbol{p})\alpha_R(\boldsymbol{p})e^{-ip\cdot x} + \epsilon u^*(-\boldsymbol{p})\beta_R^\dagger(\boldsymbol{p})e^{ip\cdot x}\right], \qquad (5.41)\end{aligned}$$

where $\alpha_R^\dagger(\boldsymbol{p})$ and $\beta_R^\dagger(\boldsymbol{p})$ respectively create massless particles (antiparticles) with helicity (the projection of the spin along the direction of motion) $h = +\frac{1}{2}$ and $h = -\frac{1}{2}$. To see this, we consider the operator on the Fock space which corresponds to a rotation by an angle θ around the axis $\boldsymbol{n}$,

$$O(\theta, \boldsymbol{n}) = e^{-i\theta \boldsymbol{J}\cdot\boldsymbol{n}}. \qquad (5.42)$$

From Eq. (5.24) we find

$$O^\dagger(\theta, \boldsymbol{n})\xi_R(x)O(\theta, \boldsymbol{n}) = e^{-i\frac{\theta\boldsymbol{\sigma}\cdot\boldsymbol{n}}{2}}\xi_R(O^{-1}x). \qquad (5.43)$$

Also,

$$\langle\Omega|\xi_R(x)\alpha_R^\dagger(\boldsymbol{p})|\Omega\rangle = \frac{u(\boldsymbol{p})e^{-ip\cdot x}}{(2\pi)^{\frac{3}{2}}}. \qquad (5.44)$$

On the other hand, using the rotation invariance of the vacuum state and choosing $\boldsymbol{n}$ as the momentum direction $\boldsymbol{p}/|\boldsymbol{p}|$, we find

$$\begin{aligned}\langle\Omega|\xi_R(x)\alpha_R^\dagger(\boldsymbol{p})|\Omega\rangle &= \langle\Omega|O^\dagger(\theta, \boldsymbol{n})\xi_R(x)O(\theta, \boldsymbol{n})O^\dagger(\theta, \boldsymbol{n})\alpha_R^\dagger(\boldsymbol{p})|\Omega\rangle \\ &= e^{ih\theta}e^{-i\frac{\theta\boldsymbol{\sigma}\cdot\boldsymbol{p}}{2|\boldsymbol{p}|}}\frac{u(\boldsymbol{p})e^{-ip\cdot x}}{(2\pi)^{\frac{3}{2}}} \\ &= e^{i(h-\frac{1}{2})\theta}\frac{u(\boldsymbol{p})e^{-ip\cdot x}}{(2\pi)^{\frac{3}{2}}}. \qquad (5.45)\end{aligned}$$

Comparing Eqs. (5.44) and (5.45) we conclude that $\alpha_R^\dagger$ creates a particle with $h = \frac{1}{2}$. In much the same way, considering the matrix element $\langle\Omega|\beta_R(\boldsymbol{p})\xi_R(x)|\Omega\rangle$ one can check that $\beta_R^\dagger$ creates particles with $h = -\frac{1}{2}$. It follows that the Lagrangian density $\mathcal{L}_{\text{Weyl}}^{(R)}$ is not invariant under the effect of space inversion, which changes the sign of h.

Finally, we observe that the Lagrangian density Eq. (5.32) is manifestly invariant under phase transformations of the fields, similar to Eq. (2.60):

$$\xi_{\mathrm{R}}(x) \to \xi'_{\mathrm{R}}(x) = e^{i\alpha}\xi_{\mathrm{R}}(x). \tag{5.46}$$

According to Eq. (2.56), the corresponding conserved current is given by

$$J^{\mu}_{\mathrm{R}} = -\xi^{\dagger}_{\mathrm{R}}\bar{\sigma}^{\mu}\xi_{\mathrm{R}}. \tag{5.47}$$

We have studied the properties of the spinor field ξ_{R} under the action of Lorentz transformations. We now discuss its behaviour under the effect of *parity inversion*, that is, the change of axis orientation in ordinary three-dimensional space. This operation is usually denoted by P. Let us consider the matrix in Eq. (5.17); under parity inversion, the parameter χ, which is related to the relative velocity of the two reference frames, changes sign, while the angle θ is unchanged. It is easy to see that such a transformation is achieved by transforming L as follows:

$$L \to L_P = \epsilon\, L^*\, \epsilon^T. \tag{5.48}$$

It is also easy to show that L_P is not equivalent to L, in the sense that it cannot be reduced to L by a transformation

$$L \to L' = U\, L\, U^{-1} \tag{5.49}$$

with U independent of θ and χ. Hence, the P-reflected image of ξ_{R} is a new spinor field, conventionally denoted by $i\xi_{\mathrm{L}}$:

$$\mathrm{P} : \xi_{\mathrm{R}}(x) \to i\xi_{\mathrm{L}}(x_P); \qquad x^0_P = x^0, \;\; \boldsymbol{x}_P = -\boldsymbol{x}. \tag{5.50}$$

that transforms as

$$\xi_{\mathrm{L}}(x) \to \xi'_{\mathrm{L}}(x) = \epsilon\, L^*\, \epsilon^T\, \xi_{\mathrm{L}}(\Lambda^{-1}(L)x). \tag{5.51}$$

Note that ξ_{L} transforms as $\epsilon\, \xi^*_{\mathrm{R}}$:

$$\epsilon\, \xi^*_{\mathrm{R}} \to \epsilon\, L^*\, \xi^*_{\mathrm{R}} = \epsilon\, L^*\, \epsilon^T\, (\epsilon\, \xi^*_{\mathrm{R}}). \tag{5.52}$$

It is easy to check that $\xi^{\dagger}_{\mathrm{L}}\, \sigma_{\mu}\, \xi_{\mathrm{L}}$ transforms as a covariant four-vector; therefore,

$$\mathcal{L}^{(L)}_{\mathrm{Weyl}} = i\xi^{\dagger}_{\mathrm{L}}\, \sigma_{\mu}\, \partial^{\mu}\xi_{\mathrm{L}} \tag{5.53}$$

is a good Lagrangian density for ξ_{L}. By the same argument employed for ξ_{R}, one can show that the particles associated to ξ_{L} have negative helicity.

The sum

$$\mathcal{L}_{\text{Weyl}} = \mathcal{L}^{(R)}_{\text{Weyl}} + \mathcal{L}^{(L)}_{\text{Weyl}} \tag{5.54}$$

is invariant under parity inversion:

$$P : i\,\xi^{\dagger}_{\text{R}}\bar{\sigma}_{\mu}\,\partial^{\mu}\xi_{\text{R}} \leftrightarrow i\,\xi^{\dagger}_{\text{L}}\sigma_{\mu}\,\partial^{\mu}\xi_{\text{L}}. \tag{5.55}$$

Assuming for $\xi_{\text{L}}(x)$ the same decomposition as in Eq. (5.41) with $u(\boldsymbol{p})$ and $v(\boldsymbol{p})$ replaced by $u(-\boldsymbol{p})$ and $v(-\boldsymbol{p})$ and $\alpha_{\text{R}}(\boldsymbol{p})$ and $\beta_{\text{R}}(\boldsymbol{p})$ replaced by $\alpha_{\text{L}}(\boldsymbol{p})$ and $\beta_{\text{L}}(\boldsymbol{p})$, that is

$$\xi_{\text{L}}(x) = \frac{1}{(2\pi)^{\frac{3}{2}}}\int d^3p[u(-\boldsymbol{p})\alpha_{\text{L}}(\boldsymbol{p})e^{-ip\cdot x} + \epsilon u^*(\boldsymbol{p})\beta^{\dagger}_{\text{L}}(\boldsymbol{p})e^{ip\cdot x}], \tag{5.56}$$

the parity reflection, in the Foch space generated by the left and right creation operators, corresponds to a unitary operator P such that $P^2 = -I$ and $P^{\dagger}\alpha_{R/L}(\boldsymbol{p})P = i\alpha_{L/R}(-\boldsymbol{p})$ up to a $\boldsymbol{p}$-independent phase factor, and $P^{\dagger}\beta_{R/L}(\boldsymbol{p})P = -i\beta_{L/R}(-\boldsymbol{p})$ with opposite phase factor. It is easy to verify that, due to our choice of the spinors u and v these equations correspond to

$$P^{\dagger}\xi_{R/L}(x)P = i\xi_{L/R}(x_P) \tag{5.57}$$

up to the above-mentioned phase factors.

A further transformation that leaves Eq. (5.54) unchanged is

$$\xi_{\text{L}}(x) \to -\epsilon\,\xi^*_{\text{R}}(x), \qquad \xi_{\text{R}}(x) \to \epsilon\,\xi^*_{\text{L}}(x). \tag{5.58}$$

Indeed, under the transformations Eq. (5.58) $\mathcal{L}_{\text{Weyl}}$ transforms as

$$\begin{aligned} & i\,\xi^T_{\text{L}}\,\epsilon^T\,\bar{\sigma}_{\mu}\,\epsilon\,\partial^{\mu}\,\xi^*_{\text{L}} + i\,\xi^T_{\text{R}}\,\epsilon^T\,\sigma_{\mu}\,\epsilon\,\partial^{\mu}\,\xi^*_{\text{R}} \\ &= -i\,\partial^{\mu}\,\xi^{\dagger}_{\text{L}}\,\epsilon^T\,\bar{\sigma}^T_{\mu}\,\epsilon\,\xi_{\text{L}} - i\,\partial^{\mu}\,\xi^{\dagger}_{\text{R}}\,\epsilon^T\,\sigma^T_{\mu}\,\epsilon\,\xi_{\text{R}} \\ &= -i\,\xi^{\dagger}_{\text{L}}\,\epsilon\bar{\sigma}^T_{\mu}\,\epsilon\,\partial^{\mu}\,\xi_{\text{L}} - i\,\xi^{\dagger}_{\text{R}}\,\epsilon\sigma^T_{\mu}\,\epsilon\,\partial^{\mu}\,\xi_{\text{R}}, \end{aligned} \tag{5.59}$$

where we have omitted total derivatives and we have taken into account the anti-commuting character of the spinor fields. Using Eq. (5.30) we get

$$\mathcal{L}_{\text{Weyl}} \to i(\xi^{\dagger}_{\text{L}}\,\sigma_{\mu}\,\partial^{\mu}\,\xi_{\text{L}} + \xi^{\dagger}_{\text{R}}\,\bar{\sigma}_{\mu}\,\partial^{\mu}\,\xi_{\text{R}}) = \mathcal{L}_{\text{Weyl}} \tag{5.60}$$

as announced. Note that, much in the same way as the transformation in Eqs. (5.50) and (5.58) is defined up to two arbitrary phase factors. This is obvious since the Weyl theory is left invariant by independent phase transformations of the spinor fields.

Now, considering the physical content of the transformations Eq. (5.58), we notice that a transformation of a field into a hermitian conjugate one corresponds to a particle-antiparticle transformation. In the present case the left-handed particle transforms into the anti-right-handed particle, which is left-handed, as it should, since Eq. (5.58) does not act on space-time. Thus we can call the action of transformation Eq. (5.58) particle-antiparticle conjugation, or *charge conjugation*, usually denoted by C. It is left as an exercise to the reader to verify that the conserved current $-(\xi_L^\dagger \bar{\sigma}_\mu \xi_L + \xi_R^\dagger \sigma_\mu \xi_R)$ changes sign under C conjugation. One can also check that both $\mathcal{L}^{(R)}_{\text{Weyl}}$ and $\mathcal{L}^{(L)}_{\text{Weyl}}$ are individually invariant under the combined action of P and C conjugations.

In much the same way as the parity reflection, the charge conjugation in the Foch space generated by the left and right creation operators corresponds to a unitary operator C such that $C^2 = I$ and

$$C^\dagger \alpha_{R/L}(\boldsymbol{p}) C = \mp \beta_{L/R}(\boldsymbol{p}); \qquad C^\dagger \beta_{R/L}(\boldsymbol{p}) C = \pm \alpha_{L/R}(\boldsymbol{p}). \tag{5.61}$$

It is easy to check that these equations correspond to

$$C^\dagger \xi_{R/L}(x) C = \pm \epsilon \xi^*_{L/R}(x) \tag{5.62}$$

up to the above phase factors.

It is worth noticing that the conventions used above, up to $\boldsymbol{p}$-independent phase factors, are such that the actions of parity inversion and charge conjugation on the Fock space correspond to local actions on the fields in space-time. If the above mentioned phase factors depended on $\boldsymbol{p}$, the action of the parity reflection and charge conjugation on the field would not be local anymore.

Now it is natural to consider the combined action of charge conjugation and parity reflection that is of CP which, forgetting phase factors, acts according to

$$C^\dagger P^\dagger \alpha_{R/L}(\boldsymbol{p}) P C = \mp i \beta_{R/L}(-\boldsymbol{p}), \qquad C^\dagger P^\dagger \beta_{R/L}(\boldsymbol{p}) P C = \mp i \alpha_{R/L}(-\boldsymbol{p}) \tag{5.63}$$

and hence

$$C^\dagger P^\dagger \xi_{R/L}(x) P C = \mp i \epsilon \xi^*_{R/L}(x_P). \tag{5.64}$$

Notice that if one multiplies $\xi_{R/L}(x)$ by the space-time independents phase factors $e^{i\varphi_{R/L}}$, the square of the same factors appear in the right-hand side of Eq. (5.64). It is also worth noticing that the charge-parity reflection acts independently on the right-handed and the left handed Fock space. For this reason it leaves invariant both $\mathcal{L}^{(R)}_{Weyl}$ and $\mathcal{L}^{(L)}_{Weyl}$.

The freedom in the spinor phase choices in the theory of fundamental interactions will be the subject of Chap. 9 of this book. It will turn out that only one phase for quarks and one for each lepton flavour are left free.

A third discrete transformation of great importance in particle physics is time reversal T. This reflection is anti-linear and exchanges initial and final states of any given process. The invariance under T conjugation is often considered as equivalent to the symmetry under CP, since a general theorem of Quantum Field Theory asserts that any covariant, dynamically stable and local theory is invariant under the combined action of P, C, and T conjugations. This invariance guarantees the identity between particle and antiparticle masses.

5.2 Mass Terms and Coupling to Scalars

In this section we study the possibility of building mass terms (i.e., bilinears in the fields without derivatives) for spinor fields. A straightforward extension is the formation of interaction terms among spinor and scalar fields.

First of all, we note that the conventional scalar product between two spinors

$$\xi_{\mathrm{R}}^{\dagger}\eta_{\mathrm{R}} = \sum_{\alpha=1}^{2}(\xi_{\mathrm{R}}^{*})_{\alpha}(\eta_{\mathrm{R}})_{\alpha} \tag{5.65}$$

is not Lorentz-invariant, because the transformation matrix L is not unitary. Instead, the quantity $\xi_{\mathrm{R}}^{T}\,\epsilon\,\eta_{\mathrm{R}}$ is invariant: using Eq. (5.25) we find

$$\xi_{\mathrm{R}}^{T}\,\epsilon\,\eta_{\mathrm{R}} \to \xi_{\mathrm{R}}^{T}\,L^{T}\epsilon L\,\eta_{\mathrm{R}} = \xi_{\mathrm{R}}^{T}\,\epsilon\,\eta_{\mathrm{R}}. \tag{5.66}$$

Similarly, one can check that $\xi_{\mathrm{L}}^{T}\epsilon\eta_{\mathrm{L}}$ is also invariant. Consider now a right-handed spinor ξ_{R} and a left-handed one, ξ_{L}. Since ξ_{L} transforms as $\epsilon\xi_{\mathrm{R}}^{*}$, it can be written as $\xi_{\mathrm{L}} = \epsilon\eta_{\mathrm{R}}^{*}$ for some right-handed spinor η_{R}. Thus,

$$\xi_{\mathrm{L}}^{\dagger}\,\xi_{\mathrm{R}} = -\eta_{\mathrm{R}}^{T}\,\epsilon\,\xi_{\mathrm{R}} \tag{5.67}$$

is also invariant.

In a generic theory, built out of the spinor and scalar fields

$$\xi_{\mathrm{R}}^{(r)}, \quad r = 1, \ldots, n_r \tag{5.68}$$

$$\xi_{\mathrm{L}}^{(l)}, \quad l = 1, \ldots, n_l \tag{5.69}$$

$$\phi^{(s)}, \quad s = 1, \ldots, n_s \tag{5.70}$$

the following terms can be included in the Lagrangian density:

$$\begin{aligned}\mathcal{L}_M = &-\sum_{rl}\left[m_{rl}\,\xi_{\mathrm{R}}^{(r)\dagger}\xi_{\mathrm{L}}^{(l)} + m_{rl}^{*}\,\xi_{\mathrm{L}}^{(l)\dagger}\xi_{\mathrm{R}}^{(r)}\right]\\ &-\frac{1}{2}\sum_{rr'}\left[M_{rr'}^{(R)}\xi_{\mathrm{R}}^{(r)T}\,\epsilon\,\xi_{\mathrm{R}}^{(r')} - M_{rr'}^{(R)*}\xi_{\mathrm{R}}^{(r')\dagger}\,\epsilon\,\xi_{\mathrm{R}}^{(r)*}\right]\\ &-\frac{1}{2}\sum_{ll'}\left[M_{ll'}^{(L)}\xi_{\mathrm{L}}^{(l)T}\,\epsilon\,\xi_{\mathrm{L}}^{(l')} - M_{ll'}^{(L)*}\xi_{\mathrm{L}}^{(l')\dagger}\,\epsilon\,\xi_{\mathrm{L}}^{(l)*}\right]\end{aligned} \tag{5.71}$$

$$\mathcal{L}_{\text{Yukawa}} = -\sum_{rls}\left[g_{rls}\phi^{(s)}\xi_{\text{R}}^{(r)\dagger}\,\xi_{\text{L}}^{(l)} + g^*_{rls}\phi^{(s)\dagger}\xi_{\text{L}}^{(l)\dagger}\,\xi_{\text{R}}^{(r)}\right]$$
$$-\sum_{rr's}\left[G^{(R)}_{rr's}\phi^{(s)}\xi_{\text{R}}^{(r)T}\,\epsilon\,\xi_{\text{R}}^{(r')} - G^{(R)*}_{rr's}\phi^{(s)\dagger}\xi_{\text{R}}^{(r')\dagger}\,\epsilon\,\xi_{\text{R}}^{(r)*}\right]$$
$$-\sum_{ll's}\left[G^{(L)}_{ll's}\phi^{(s)}\xi_{\text{L}}^{(l)T}\,\epsilon\,\xi_{\text{L}}^{(l')} - G^{(L)*}_{ll's}\phi^{(s)\dagger}\xi_{\text{L}}^{(l')\dagger}\,\epsilon\,\xi_{\text{L}}^{(l)*}\right]. \quad (5.72)$$

Spinor fields have the dimension of an energy to the power 3/2, while scalar fields have the dimension of energy. Hence, since the Lagrangian density has the dimension of an energy to the fourth power, the constants $M^{(R)}$, $M^{(L)}$, m have the dimension of an energy, and $G^{(R)}$, $G^{(L)}$, g are dimensionless.

The term $\mathcal{L}_M$ contains mass terms for the spinor fields. The first line in Eq. (5.71) is invariant under multiplication of all spinor fields by a common phase factor, and is usually called a *Dirac mass term*. The remaining two terms do not share the same property, and are called *Majorana mass terms*. The mass matrices $M^{(R)}$ and $M^{(L)}$ are symmetric, as one can check taking into account the anti-commutation relations of spinor fields mentioned in Sect. 5.1 (which is related to Fermi–Dirac statistics) and the antisymmetry of ϵ. Similar considerations hold for $\mathcal{L}_{\text{Yukawa}}$; spinor-spinor-scalar interaction terms are usually referred to as*Yukawa couplings*.

Let us concentrate on mass terms. To simplify notations, we collect all left-handed spinors $\xi_{\text{L}}^{(l)}$ and $\epsilon\xi_{\text{R}}^{(r)*}$ in an $(n_l + n_r)$-component spinor Ξ. Then, we may write Eq. (5.71) in the form

$$\mathcal{L}_M = -\frac{1}{2}\sum_{\alpha,\beta}\left[\mathcal{M}_{\alpha\beta}\Xi^T_\alpha\,\epsilon\,\Xi_\beta - \mathcal{M}^*_{\beta\alpha}\Xi^\dagger_\alpha\,\epsilon\,\Xi^*_\beta\right], \quad (5.73)$$

where $\mathcal{M}$ is a complex symmetric matrix, with

$$\mathcal{M}_{ll'} = M^{(L)}_{ll'} \qquad \mathcal{M}_{rr'} = -M^{(R)*}_{rr'} \qquad \mathcal{M}_{lr} = m_{lr}. \quad (5.74)$$

A complex symmetric matrix can always be written in the form

$$\mathcal{M} = \mathcal{U}^T\hat{\mathcal{M}}\mathcal{U}, \quad (5.75)$$

with$\mathcal{U}$ unitary and $\hat{\mathcal{M}}$ real and diagonal (the number of real parameters that determine an $N\times N$ matrix is $N(N+1)$ if the matrix is symmetric, N^2 if it is unitary, N if it is real and diagonal). On the other hand, a unitary transformation on Ξ,

$$\Xi \to \mathcal{U}\,\Xi; \qquad \mathcal{U}^\dagger\mathcal{U} = I \quad (5.76)$$

does not affect kinetic terms in the canonical form,

$$\mathcal{L}_k = \Xi^\dagger\sigma_\mu\partial^\mu\Xi, \quad (5.77)$$

and brings $\mathcal{L}_M$ in diagonal form:

$$\mathcal{L}_M = -\frac{1}{2}\sum_\alpha \hat{\mathcal{M}}_{\alpha\alpha}\left(\Xi_\alpha^T\,\epsilon\,\Xi_\alpha - \Xi_\alpha^\dagger\,\epsilon\,\Xi_\alpha^*\right). \tag{5.78}$$

We have shown that $\mathcal{L}_M$ can always be written as a sum of Majorana mass terms, with real and positive coefficients, without any Dirac mass terms. At first sight, this is a surprising result, since we know that only Dirac mass terms are compatible with invariance under phase rotations. However, it is easy to show that each Dirac mass term in the original Lagrangian corresponds to a pair of degenerate eigenvalues in $\mathcal{L}_M$ written in the canonical Majorana form, Eq. (5.78); a rotation in the subspace of the two degenerate spinor fields appears as a phase transformation.

Let us consider the case when Ξ has only two components $\xi_\mathrm{L}, \epsilon\xi_\mathrm{R}^*$. The mass matrix is given by

$$\mathcal{M} = \begin{pmatrix} a & b \\ b & c \end{pmatrix} \tag{5.79}$$

with a, b, c complex. The squared masses are given by the eigenvalues of

$$\mathcal{M}^\dagger\mathcal{M} = \begin{pmatrix} a^* & b^* \\ b^* & c^* \end{pmatrix}\begin{pmatrix} a & b \\ b & c \end{pmatrix} = \begin{pmatrix} |a|^2+|b|^2 & a^*b+b^*c \\ ab^*+bc^* & |b|^2+|c|^2 \end{pmatrix}, \tag{5.80}$$

that is,

$$m_\pm^2 = \frac{1}{2}\Big[|a|^2+|c|^2+2|b|^2 \\ \pm\sqrt{\left(|a|^2-|c|^2\right)^2+4\left[|b|^2\left(|a|^2+|c|^2\right)+2\mathrm{Re}\,\left(a^*c^*b^2\right)\right]}\Big]. \tag{5.81}$$

An interesting limit of Eq. (5.81) is the case $c = 0$, $|b| \ll |a|$. In this case we find

$$m_\pm^2 = \frac{1}{2}\left[|a|^2+2|b|^2 \pm \sqrt{|a|^4+4|b|^2|a|^2}\right]. \tag{5.82}$$

and therefore, to first order in $|b/a|^2$,

$$m_+^2 \simeq |a|^2; \qquad m_-^2 \simeq |b|^2\left|\frac{b}{a}\right|^2. \tag{5.83}$$

This is the so called *see-saw* mechanism, which is employed to explain the small values of neutrino masses, as we shall see in Chap. 13.

Under CP conjugation,

$$\Xi_\alpha(x) \to i\epsilon\,\Xi_\alpha^*(x_P), \tag{5.84}$$

which is obtained from Eq. (5.64). This leaves Eqs. (5.78) and (5.54) invariant (up to total derivatives). Coupling terms are left invariant provided a transformation as in Eq. (5.76) exists, such that all parameters are made real.

The action of parity reflection on the multispinor Ξ is more involved. Indeed, parity reflection acts differently on the components of Ξ which originally were left-handed spinor components, and on those which were complex conjugate right-handed ones, transforming elements of the first kind into complex conjugate elements of the second one and viceversa. Therefore, if the theory is parity invariant, the number of Ξ components must be even and the eighenvalues of the mass matrix $\mathcal{M}$ associated with parity conjugate elements must coincide. As mentioned above this degeneracy of $\mathcal{M}$ identifies $\mathcal{L}_M$ with a Dirac mass term.

Chapter 6
Gauge Symmetries

6.1 Electrodynamics

The simplest example of a phenomenological application of the theory of spinor fields is electrodynamics, originally formulated by Dirac to provide a relativistic description of electrons and photons. The theory is based on a pair of spinor fields of opposite chiralities, ξ_R and ξ_L; invariance under phase multiplication of the spinor fields is assumed, and the corresponding conserved current is identified with the electromagnetic current. No scalar field is present. Under these assumptions, the free-field Lagrangian is given by

$$\mathcal{L}_0 = i\xi_R^\dagger \,\bar{\sigma}_\mu \,\partial^\mu \,\xi_R + i\xi_L^\dagger \,\sigma_\mu \,\partial^\mu \,\xi_L - m\left(\xi_R^\dagger \,\xi_L + \xi_L^\dagger \,\xi_R\right). \tag{6.1}$$

$\mathcal{L}_0$ is manifestly invariant under the phase transformations

$$\xi_R(x) \to e^{ie\Lambda}\,\xi_R(x) \tag{6.2}$$

$$\xi_L(x) \to e^{ie\Lambda}\,\xi_L(x), \tag{6.3}$$

where e, Λ are real constants. Notice that we have inserted a Dirac mass term, which is allowed by the assumed invariance properties, while Majorana mass terms are not, as mentioned in Sect. 5.2. The conserved current is given by

$$J_\mu = -e\left(\xi_R^\dagger \,\bar{\sigma}_\mu \,\xi_R + \xi_L^\dagger \,\sigma_\mu \,\xi_L\right); \qquad \partial^\mu J_\mu = 0, \tag{6.4}$$

and the constant e plays the role of elementary charge.

It is known from classical physics that electromagnetic fields interact with matter through the electric current density $e\boldsymbol{J}$ and the charge density $e\rho$; the interaction energy has the form

$$H_I = \frac{e}{c}\int d^3r\,\left(\rho\,\Phi_{\text{em}} - \boldsymbol{J}\cdot\boldsymbol{A}\right), \tag{6.5}$$

C. M. Becchi and G. Ridolfi, *An Introduction to Relativistic Processes and the Standard Model of Electroweak Interactions*, UNITEXT for Physics,
DOI: 10.1007/978-3-319-06130-6_6,

where e is the elementary charge, $\Phi_{\rm em}$ the electrostatic potential, $\boldsymbol{A}$ the vector potential, ρ the position density of charged particles, or the charge density divided by the elementary charge, and $\boldsymbol{J}$ the current density. In the covariant formalism, one recognizes that $e(c\rho, \boldsymbol{J})$ form a current four-vector J_μ, and $\left(\frac{\Phi_{\rm em}}{c}, \boldsymbol{A}\right)$ a potential four-vector A^μ; hence, in our units,

$$H_I = \int d^3r\, A^\mu J_\mu. \tag{6.6}$$

We will therefore take

$$\mathcal{L}_I = -A^\mu J_\mu = e\, A^\mu \left(\xi_R^\dagger\, \bar{\sigma}_\mu\, \xi_R + \xi_L^\dagger\, \sigma_\mu\, \xi_L\right) \tag{6.7}$$

as the Lagrangian density for the electromagnetic interactions. Thus,

$$\mathcal{L}_{\rm QED} = i\xi_R^\dagger\, \bar{\sigma}_\mu(\partial^\mu - ieA^\mu)\xi_R + i\xi_L^\dagger\, \sigma_\mu(\partial^\mu - ieA^\mu)\xi_L - m\left(\xi_R^\dagger\, \xi_L + \xi_L^\dagger\, \xi_R\right). \tag{6.8}$$

The theory is invariant with respect to both parity inversion P and charge conjugation C, assuming that the vector field A_μ changes sign under charge conjugation.

Notations are remarkably simplified by the following, universally adopted, definitions. The two spinors are paired into a single four-component spinor

$$\psi(x) \equiv \begin{pmatrix} \xi_R(x) \\ \xi_L(x) \end{pmatrix} \equiv \begin{pmatrix} \xi_{R1}(x) \\ \xi_{R2}(x) \\ \xi_{L1}(x) \\ \xi_{L2}(x) \end{pmatrix}. \tag{6.9}$$

Next, one defines the 4×4 matrices

$$\gamma_\mu \equiv \begin{pmatrix} 0 & \sigma_\mu \\ \bar{\sigma}_\mu & 0 \end{pmatrix}, \quad \gamma_5 \equiv \begin{pmatrix} I & 0 \\ 0 & -I \end{pmatrix} = -i\gamma_0\gamma_1\gamma_2\gamma_3 \tag{6.10}$$

and the Dirac-conjugate spinor

$$\bar{\psi}(x) \equiv \left(\xi_L^\dagger(x)\; \xi_R^\dagger(x)\right) = \psi^\dagger \gamma_0. \tag{6.11}$$

Through the matrix γ_5, it is possible to define projection operators, that project the left and right components out of the four-spinor ψ:

$$\frac{1+\gamma_5}{2}\psi = \begin{pmatrix} \xi_R \\ 0 \end{pmatrix}; \quad \frac{1-\gamma_5}{2}\psi = \begin{pmatrix} 0 \\ \xi_L \end{pmatrix}. \tag{6.12}$$

The notation

$$\not{q} \equiv \gamma_\mu q^\mu \tag{6.13}$$

is often used. With these definitions, the Lagrangian density of electrodynamics becomes

$$\mathcal{L}_{\text{QED}} = i\bar{\psi}\gamma_\mu\left(\partial^\mu - ieA^\mu\right)\psi - m\bar{\psi}\psi. \tag{6.14}$$

An even simpler expression is obtained introducing the *covariant derivative* of the field ψ as

$$D_\mu\psi(x) \equiv \partial_\mu\psi(x) - ieA_\mu(x)\psi(x). \tag{6.15}$$

The new, important fact that arises after introduction of the interaction with the electromagnetic field is that the Lagrangian density

$$\mathcal{L}_{\text{QED}} = i\bar{\psi}\gamma_\mu D^\mu\psi - m\bar{\psi}\psi \tag{6.16}$$

is now invariant with respect to the transformations

$$\psi(x) \to e^{+ie\Lambda(x)}\psi(x) \tag{6.17}$$

$$\bar{\psi}(x) \to e^{-ie\Lambda(x)}\bar{\psi}(x) \tag{6.18}$$

$$A^\mu(x) \to A^\mu(x) + \partial^\mu\Lambda(x), \tag{6.19}$$

that are a generalization of the transformations in Eq. (2.60) to the case $\Lambda = \Lambda(x)$. For historical reasons, the transformations (6.17–6.19) are called *gauge transformations*.

It is now clear why D^μ is called a covariant derivative: from Eq. (6.17) we see that the transformation of the ordinary partial derivative of ψ,

$$\partial_\mu\psi(x) \to \partial_\mu\left(e^{ie\Lambda(x)}\psi(x)\right) = e^{ie\Lambda(x)}\partial_\mu\psi(x) + ie\partial_\mu\Lambda(x)e^{ie\Lambda(x)}\psi(x) \tag{6.20}$$

is not just a multiplication by a phase factor. This is instead true for $D_\mu\psi$, since the second term in Eq. (6.15) transforms as

$$-ieA_\mu(x)\psi(x) \to -ieA_\mu(x)e^{ie\Lambda(x)}\psi(x) - ie\partial_\mu\Lambda(x)e^{ie\Lambda(x)}\psi(x), \tag{6.21}$$

thus compensating the term proportional to $\partial_\mu\Lambda$ in Eq. (6.20). We have therefore established that

$$D_\mu\psi(x) \to e^{ie\Lambda(x)}D_\mu\psi(x). \tag{6.22}$$

It is immediate to show that the same result extends to multiple covariant derivatives:

$$D_{\mu_1}\ldots D_{\mu_n}\psi(x) \to e^{ie\Lambda(x)}D_{\mu_1}\ldots D_{\mu_n}\psi(x) \tag{6.23}$$

for any value of n.

The above construction is the simplest application of the *principle of gauge invariance*. It can be shown on general grounds that any field theory that involves four-vector fields must possess a gauge invariance under transformations analogous to Eqs. (6.17–6.19). Gauge invariance is needed in order to eliminate the effect of those components of the vector field that do not correspond to physical degrees of freedom. Such components are necessarily present, since the vector field has four components, while the associated spin-1 particles have at most three degrees of freedom. An explicit example of this cancellation will be shown in Chap. 8 in the context of the Higgs model.

In order to complete the Lagrangian density of electrodynamics, we should add a term for the electromagnetic field. The energy density of electric and magnetic fields is given by $\frac{\boldsymbol{E}^2+\boldsymbol{B}^2}{2}$; the term proportional to $\boldsymbol{E}^2$ can be interpreted as a 'kinetic' term, because the electric field $\boldsymbol{E}$ is linear in the time derivative of the vector potential $\boldsymbol{A}$, while the magnetic field $\boldsymbol{B}$ is given by spatial derivatives (the curl of $\boldsymbol{A}$). Hence, we conclude that the electromagnetic Lagrangian density is

$$\mathcal{L}_{\rm em} = \frac{\boldsymbol{E}^2 - \boldsymbol{B}^2}{2}. \tag{6.24}$$

This can be written in an explicitly covariant form by means of the *field tensor*

$$F_{\mu\nu} = \partial_\mu A_\nu - \partial_\nu A_\mu. \tag{6.25}$$

We find

$$\mathcal{L}_{\rm em} = -\frac{1}{4} F^{\mu\nu} F_{\mu\nu}. \tag{6.26}$$

The tensor $F_{\mu\nu}$ is invariant under gauge transformations, as one can verify directly using Eq. (6.19). Alternatively, one may observe that

$$F_{\mu\nu}(x)\psi(x) = \frac{i}{e}(D_\mu D_\nu - D_\nu D_\mu)\psi(x) \equiv \frac{i}{e}[D_\mu, D_\nu]\psi(x), \tag{6.27}$$

from which it is immediately clear that $F_{\mu\nu}$, and therefore $\mathcal{L}_{\rm em}$, are gauge-invariant. We should stress that $\mathcal{L}_{\rm em}$ is the only possible term that depends on A^μ and its first derivatives, compatible with gauge and Lorentz invariance, C and P invariance, and renormalizability. In particular, gauge invariance in the form of Eqs. (6.17), (6.18) and (6.19) excludes the possibility of introducing a term $\mu^2 A_\mu A^\mu$, that would correspond to a non-zero value for the photon mass.[1]

[1] As a matter of fact, there is an alternative formulation of gauge invariance, associated with the Higgs mechanism, which we shall describe in Sect. 8.2. As remarked at the end of this section, in a particular limit, which is only allowed in an abelian gauge theory, such as QED, the Higgs mechanism corresponds to the introduction of a mass term for the photon. The resulting theory is Stueckelberg's massive QED.

The electromagnetic Lagrangian density is therefore

$$\mathcal{L}_{\text{QED}} = i\bar{\psi}\gamma^\mu D_\mu \psi - m\bar{\psi}\psi - \frac{1}{4}F^{\mu\nu}F_{\mu\nu}. \tag{6.28}$$

The equations of motion are immediately derived. We find

$$i\gamma^\mu \partial_\mu \psi - m\psi = -e\,\gamma^\mu A_\mu \psi \tag{6.29}$$

$$\partial^2 A^\mu - \partial^\mu \partial A = -J^\mu. \tag{6.30}$$

Next, we must determine the Green functions for both equations. The Green function $S(x)$ for the spinor field is defined by

$$\left(i\gamma^\mu \partial_\mu - m\right) S(x) = \delta(x). \tag{6.31}$$

The calculation of $S(x)$ is greatly simplified by the algebraic properties of the γ matrices,

$$\{\gamma^\mu, \gamma^\nu\} = 2I g^{\mu\nu}, \qquad \text{Tr}(\gamma^\mu \gamma^\nu) = 4g^{\mu\nu}. \tag{6.32}$$

We multiply both sides of Eq. (6.31) with $i\gamma^\mu\partial_\mu + m$ on the left:

$$\left(i\gamma^\mu \partial_\mu + m\right)\left(i\gamma^\nu \partial_\nu - m\right) S(x) = \left(i\gamma^\mu \partial_\mu + m\right)\delta(x), \tag{6.33}$$

and we observe that, because of Eq. (6.32),

$$\left(i\gamma^\mu \partial_\mu + m\right)\left(i\gamma^\nu \partial_\nu - m\right) = -\left(\partial^2 + m^2\right). \tag{6.34}$$

Hence,

$$-\left(\partial^2 + m^2\right) S(x) = \left(i\gamma^\mu \partial_\mu + m\right)\delta(x), \tag{6.35}$$

and comparing with Eqs. (3.100 and 3.102) we find

$$S(x) = -\left(i\gamma^\mu \partial_\mu + m\right)\Delta(x) = \int \frac{d^4k}{(2\pi)^4}\frac{\gamma_\mu k^\mu + m}{k^2 - m^2 + i\epsilon}e^{-ik\cdot x}, \tag{6.36}$$

where $\Delta(x)$ is the Green function for the scalar field, Eq. (3.100).

The natural definition for the Green function for A^μ would be

$$\partial^2 \Delta^{\mu\nu}(x) - \partial^\mu \partial_\lambda \Delta^{\lambda\nu}(x) = -g^{\mu\nu}\delta^4(x). \tag{6.37}$$

This equation, however, has no solution. This is easily seen by taking the four-dimensional Fourier transform of Eq. (6.37), and observing that the operator $k^2 g^{\mu\lambda} - k^\mu k^\lambda$ has no inverse, since k_λ is an eigenvector with vanishing eigenvalue. This difficulty is related to the fact that Maxwell's equations do not determine the vector potential: they must therefore be solved with a supplementary prescription. The choice of this prescription has been one of the central problems in theoretical physics during the last century; it is connected with the axioms of quantum mechanics, and in particular with the conservation of probability. However, in the semi-classical approximation of electrodynamics this problem has a simple solution, that will be presented here in a further simplified version. We introduce an extra (*gauge fixing*) term in the Lagrangian density, namely

$$\mathcal{L}_{\mathrm{gf}} = -\frac{1}{2}\left(\partial_\mu A^\mu\right)^2 . \tag{6.38}$$

The Lagrangian density becomes

$$\mathcal{L}_{\mathrm{QED}} = i\bar{\psi}\gamma^\mu D_\mu \psi - m\bar{\psi}\psi - \frac{1}{2}\partial_\mu A_\nu \partial^\mu A^\nu + \frac{1}{2}\partial_\mu A_\nu \partial^\nu A^\mu - \frac{1}{2}\left(\partial_\mu A^\mu\right)^2 . \tag{6.39}$$

The last two terms can be rewritten as

$$\frac{1}{2}\partial_\mu A_\nu \partial^\nu A^\mu - \frac{1}{2}\left(\partial_\mu A^\mu\right)^2 = \frac{1}{2}\partial_\mu\left(A_\nu \partial^\nu A^\mu - A^\mu \partial_\nu A^\nu\right), \tag{6.40}$$

which is a four-divergence. By the Gauss-Green theorem, this contributes to the action functional only through surface terms, and therefore has no effect on the equations of motion. We may therefore omit it altogether, and write the QED Lagrangian in its final form,

$$\mathcal{L}_{\mathrm{QED}} = i\bar{\psi}\gamma^\mu D_\mu \psi - m\bar{\psi}\psi - \frac{1}{2}\partial_\mu A_\nu \partial^\mu A^\nu . \tag{6.41}$$

The equations of motion for the vector field A^μ are now

$$\partial^2 A^\mu = -J^\mu . \tag{6.42}$$

This implies that, whenever the current J^μ is conserved, $\partial . \mathrm{A}$ is a free field and hence does not participate in physical transition processes. The Green function corresponding to Eq. (6.42) is the solution of

$$\partial^2 \Delta^{\mu\nu}(x) = -g^{\mu\nu}\,\delta(x), \tag{6.43}$$

and is immediately computed using Eq. (3.100) with $m = 0$; we find

$$\Delta^{\mu\nu}(x) = g^{\mu\nu}\int \frac{d^4k}{(2\pi)^4}\frac{e^{-ik\cdot x}}{k^2 + i\epsilon} . \tag{6.44}$$

The presence of $g^{\mu\nu}$ in the photon propagator implies that the sign of the propagator of the time-like component is opposite to that of the space-like components, which is the same as for a massless scalar field. This corresponds to the fact that time-like polarization states have negative norm, as we shall see in Sect. 8.3. This is physically relevant, since the norm of states has a probabilistic interpretation: the presence of negative norms violates the conservation of probability, or equivalently the unitarity of the $\mathcal{S}$-matrix. We are going to see in a moment that time-like polarization states do not participate in the scattering process in the semi-classical approximation.

We are now able to work out the Feynman rules for electrodynamics. The Dirac field ψ is a charged field, and the corresponding lines have an orientation. The propagator $S(x-y)$, Eq. (6.36), is a matrix in the space of Dirac indices; the line index corresponds to the field $\psi(x)$, and the column index to $\bar{\psi}(y)$. The photon propagator is given in Eq. (6.44). Finally, there is only one interaction term in the Lagrangian density (6.41), namely

$$e A_\mu \bar{\psi} \gamma^\mu \psi. \tag{6.45}$$

The commonly adopted graphical symbols are

$$\leftrightarrow \quad \frac{(\gamma^\mu q_\mu + m)^\alpha_\beta}{q^2 - m^2 + i\epsilon} \tag{6.46}$$

for the spinor propagator, with the momentum q flowing in the direction of the arrow;

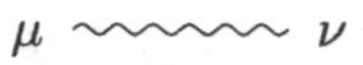

$$\leftrightarrow \quad \frac{g^{\mu\nu}}{k^2 + i\epsilon} \tag{6.47}$$

for the photon propagator, and

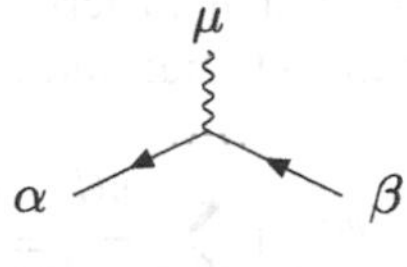

$$\leftrightarrow \quad e\,(\gamma^\mu)^\alpha_\beta \tag{6.48}$$

for the vertex factor. Note that each endpoint of a fermion line carries a spinor index (α and β in the figure), and that each endpoint of a photon line carries a space-time index (μ and ν in the figure.)

Finally, we need the expressions of the asymptotic vector and spinor fields for given initial and final states of a transition. The asymptotic vector field can be obtained from the asymptotic expression for the scalar field, Eq. (3.89), modified in order to account for the polarization states of vector particles. For each plane wave with momentum p_μ we introduce a complex polarization four-vector ϵ_μ, that depends on the helicity state λ of the asymptotic particle. The asymptotic vector field turns out to be

$$A^{(\text{as})}_\mu(x) = \frac{(\sqrt{4\pi\delta})^{3/2}}{(2\pi)^{3/2}} \left[\sum_i \epsilon^{(i)}_\mu(p_i, \lambda_i) \frac{e^{-ip_i \cdot x}}{\sqrt{2E_{p_i}}} + \sum_f \epsilon^{(f)*}_\mu(k_f, \lambda_f) \frac{e^{ik_f \cdot x}}{\sqrt{2E_{k_f}}} \right]. \tag{6.49}$$

A generic polarization vector $\epsilon^\mu(p,\lambda)$ can be decomposed as

$$\epsilon^\mu = \epsilon_T^\mu + a p^\mu + b \bar{p}^\mu, \tag{6.50}$$

where

$$\bar{p} = (p^0, -\boldsymbol{p}); \quad \epsilon_T = (0, \boldsymbol{\epsilon}_T); \quad \boldsymbol{p} \cdot \boldsymbol{\epsilon}_T = 0. \tag{6.51}$$

The asymptotic value of ∂A is therefore proportional to b. Since, as shown above, ∂A is a free field, and hence is does not participate in the scattering process, one can choose $b = 0$ without loss of generality. The longitudinal polarization component ap^μ, on the other hand, may be eliminated by a suitable gauge transformation: this is possible, because even after the introduction of the gauge fixing term $\mathcal{L}_{\text{gf}}$ the Lagrangian density has a residual gauge invariance with respect to transformations $A^\mu \to A^\mu + \partial^\mu \Lambda$ with $\partial^2 \Lambda = 0$. We conclude that the particles associated with the field A^μ can only have transverse polarization states.

The transverse polarization vectors must be normalized as

$$\epsilon_T{}^*_\mu(p,\lambda)\, \epsilon_T^\mu(p,\lambda) = -1, \tag{6.52}$$

and it is easy to show that the sum over physical photon polarizations can be performed using

$$\sum_\lambda \epsilon_\mu(p,\lambda)\, \epsilon_\nu^*(p,\lambda) = -g_{\mu\nu} + \frac{p_\mu \bar{p}_\nu + \bar{p}_\mu p_\nu}{p \cdot \bar{p}}. \tag{6.53}$$

Similarly, the asymptotic spinor field is written in terms of coefficients that describe the spin state of particles in the initial and final states:

$$\psi^{(\text{as})}(x) = \frac{(\sqrt{4\pi\delta})^{3/2}}{(2\pi)^{3/2}} \left[\sum_i \frac{u(p_i,\lambda_i)}{\sqrt{2E_{p_i}}} e^{-ip_i\cdot x} + \sum_f \frac{v(k'_f,\lambda'_f)}{\sqrt{2E_{k'_f}}} e^{ik'_f\cdot x} \right] \tag{6.54}$$

$$\bar{\psi}^{(\text{as})}(x) = \frac{(\sqrt{4\pi\delta})^{3/2}}{(2\pi)^{3/2}} \left[\sum_i \frac{\bar{v}(p'_i,\lambda'_i)}{\sqrt{2E_{p'_i}}} e^{-ip'_i\cdot x} + \sum_f \frac{\bar{u}(k_f,\lambda_f)}{\sqrt{2E_{k_f}}} e^{ik_f\cdot x} \right] \tag{6.55}$$

where

- p_i, λ_i are momenta and helicities of particles in the initial state
- k'_f, λ'_f are momenta and helicities of antiparticles in the final state
- p'_i, λ'_i are momenta and helicities of antiparticles in the initial state
- k_f, λ_f are momenta and helicities of particles in the final state.

Note that $\bar{\psi}^{(\text{as})}$ is *not* the Dirac conjugate of $\psi^{(\text{as})}$, since they depend on different physical parameters. This happens because the weak limit defined in Sect. 3.3 does not preserve the hermiticity properties of the fields. The spinors u and v obey the constraints

$$(\gamma_\mu p^\mu - m)\, u(p,\lambda) = 0 \tag{6.56}$$
$$(\gamma_\mu p^\mu + m)\, v(p,\lambda) = 0 \tag{6.57}$$

as a consequence of the equations of motion for the free Dirac field. Consequently, the second term in the polarization sum Eq. (6.53) does not contribute to the transition probability, and can be omitted altogether. This a general result in QED.

Due to the factor $1/\sqrt{2E_p}$ in Eqs. (6.54, 6.55), u and v are normalized as

$$u^\dagger(p,\lambda)\, u(p,\lambda') = v^\dagger(p,\lambda)\, v(p,\lambda') = 2E_p\, \delta_{\lambda\lambda'}; \qquad v^\dagger(\bar{p},\lambda)\, u(p,\lambda') = 0, \tag{6.58}$$

where $\bar{p} = (p^0, -\boldsymbol{p})$. In the case $m \neq 0$, the above normalization can be translated into the invariant form

$$\bar{u}(p,\lambda)\, u(p,\lambda') = 2m\, \delta_{\lambda\lambda'} \tag{6.59}$$
$$\bar{v}(p,\lambda)\, v(p,\lambda') = -2m\, \delta_{\lambda\lambda'} \tag{6.60}$$
$$\bar{v}(p,\lambda)\, u(p,\lambda') = 0. \tag{6.61}$$

These relations imply

$$\sum_\lambda u(p,\lambda)\bar{u}(p,\lambda) = \gamma_\mu p^\mu + m \tag{6.62}$$
$$\sum_\lambda v(p,\lambda)\bar{v}(p,\lambda) = \gamma_\mu p^\mu - m. \tag{6.63}$$

6.2 A Sample Calculation: Compton Scattering

As an example, we compute the cross section for the process

$$e^-(p) + \gamma(q,\epsilon) \to e^-(p') + \gamma(q',\epsilon'), \tag{6.64}$$

usually called Compton scattering (the four-momenta of the particles involved are indicated in brackets). The invariant amplitude is obtained from the diagrams

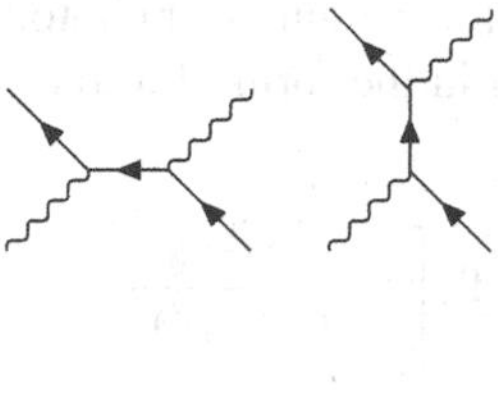

(6.65)

We find

$$\mathcal{M} = \mathcal{M}_{\alpha\beta}\, \epsilon^{\alpha}\, \epsilon'^{*\beta} \tag{6.66}$$
$$\mathcal{M}_{\alpha\beta} = e^2 \bar{u}(p') \left[\gamma_\alpha \frac{\not{p}' - \not{q} + m}{(q-p')^2 - m^2} \gamma_\beta + \gamma_\beta \frac{\not{p}' + \not{q}' + m}{(q'+p')^2 - m^2} \gamma_\alpha \right] u(p),$$

where ϵ and ϵ' are the polarization vectors of the photons in the initial and final states:

$$\epsilon \cdot q = \epsilon' \cdot q' = 0. \tag{6.67}$$

The second term in the amplitude is obtained from the first one by *crossing*, that is, by replacing $q \to -q'$ and $\epsilon \to \epsilon'^*$. Observe that

$$q^{\alpha}\, \mathcal{M}_{\alpha\beta} = q'^{\beta}\, \mathcal{M}_{\alpha\beta} = 0 \tag{6.68}$$

as a consequence of the equations of motion for $u(p)$ and $u(p')$. Therefore the polarization states with $\epsilon^{\mu} \sim p^{\mu}$ do not couple and the second term in the right-hand side of Eq. (6.53) can be omitted.

The relations in Eq. (6.68) are particular cases of a class of identities, usually called Ward identities, obeyed by physical amplitudes as a consequence of gauge invariance.

The complex conjugate amplitude is easily computed with the help of the results of Appendix F; we find

$$\mathcal{M}^*_{\alpha\beta} = e^2 \bar{u}(p) \left[\gamma_\beta \frac{\not{p}' - \not{q} + m}{(q-p')^2 - m^2} \gamma_\alpha + \gamma_\alpha \frac{\not{p}' + \not{q}' + m}{(q'+p')^2 - m^2} \gamma_\beta \right] u(p'). \tag{6.69}$$

The calculation simplifies considerably in the high-energy limit, in which the electron mass can be neglected. In this limit, we find

$$\begin{aligned}\sum_{\mathrm{pol}_\gamma} \mathcal{M}^* \mathcal{M} = {} & e^4 \bar{u}(p') \left[\gamma_\alpha \frac{\not{p}' - \not{q}}{(q-p')^2} \gamma_\beta + \gamma_\beta \frac{\not{p}' + \not{q}'}{(q'+p')^2} \gamma_\alpha \right] u(p) \\ & \times \bar{u}(p) \left[\gamma^\beta \frac{\not{p}' - \not{q}}{(q-p')^2} \gamma^\alpha + \gamma^\alpha \frac{\not{p}' + \not{q}'}{(q'+p')^2} \gamma^\beta \right] u(p'),\end{aligned} \tag{6.70}$$

where we have used Eq. (6.53) for the sum over photon helicity states, and Eq. (6.68). Equation (6.70) can be written in the form of a trace over spinor indices by means of Eqs. (6.62, 6.63):

$$\begin{aligned}\sum_{\mathrm{pol}_\gamma, pol_e} \mathcal{M}^* \mathcal{M} = e^4\, \mathrm{Tr} & \left[\gamma_\alpha \frac{\not{p}' - \not{q}}{(q-p')^2} \gamma_\beta + \gamma_\beta \frac{\not{p}' + \not{q}'}{(q'+p')^2} \gamma_\alpha \right] \not{p} \\ & \left[\gamma^\beta \frac{\not{p}' - \not{q}}{(q-p')^2} \gamma^\alpha + \gamma^\alpha \frac{\not{p}' + \not{q}'}{(q'+p')^2} \gamma^\beta \right] \not{p}'.\end{aligned} \tag{6.71}$$

The trace can be computed by means of the results of Appendix F, Eqs. (F.3–F.5) and (F.8). For example,

$$\begin{aligned}\mathrm{Tr}\, \gamma_\alpha\, (\not{p}' - \not{q})\, \gamma_\beta\, \not{p}\, \gamma^\beta\, (\not{p}' - \not{q})\, \gamma^\alpha\, \not{p}' &= 4\, (p' - q)_\mu\, p_\rho\, (p' - q)_\nu\, p'_\sigma \mathrm{Tr}\, \gamma^\mu\, \gamma^\rho\, \gamma^\nu\, \gamma^\sigma \\ &= 32\, p \cdot q'\, (p \cdot p' + p \cdot q'),\end{aligned} \tag{6.72}$$

where we have repeatedly used the mass-shell conditions and the momentum conservation relation. A straightforward calculation leads to

$$\sum_{\mathrm{pol}} \mathcal{M}^* \mathcal{M} = -8e^4 \left(\frac{t}{s} + \frac{s}{t} \right), \tag{6.73}$$

where we have defined

$$s = (p + q)^2; \qquad t = (p - q')^2. \tag{6.74}$$

In the center-of-mass frame we have

$$t = -\frac{s}{2}(1 - \cos\theta), \tag{6.75}$$

where θ is the photon scattering angle. The unpolarized differential cross section, averaged over the four initial polarizations, is now immediately obtained:

$$\begin{aligned} d\sigma &= \frac{1}{4}\frac{1}{2s} \sum_{\mathrm{pol}} \mathcal{M}^* \mathcal{M}\, d\phi_2(p + q; p', q') \\ &= \frac{e^4}{16\pi s} \frac{1 + \sin^4 \frac{\theta}{2}}{\sin^2 \frac{\theta}{2}}\, d\cos\theta, \end{aligned} \tag{6.76}$$

where we have used the expression Eq. (4.25) for the two-body invariant phase space.

6.3 Non-commutative Charges: The Yang-Mills Theory

As an obvious generalization of the symmetry transformations that characterize quantum electrodynamics, we may consider a group of transformations that act on fields in a multi-dimensional isotopic space ϕ_i, $i = 1, \ldots, N$. For definiteness, we consider the transformations

$$\phi_i \to U_{ij}(\alpha)\phi_j, \tag{6.77}$$

where U is an $N \times N$ unitary matrix with unit determinant. The matrices U do not commute with one another; this is why Eq. (6.77) is called a *non-abelian* set of transformations, as opposed to the case of multiplication by a phase factor, which is an *abelian* (or commutative) operation. The set of matrices U is a particular representation (called the *fundamental representation*) of the group $SU(N)$. Any such matrix can be written as

$$U = U(\alpha) = \exp(ig\alpha^A t^A), \tag{6.78}$$

where the t^A, $A = 1, \ldots, N^2 - 1$ are a basis in the linear space of hermitian traceless matrices, the *generators*, and α^A are real constants. The generators can always be chosen so that

$$\mathrm{Tr}\, t^A t^B = T_F \delta^{AB}, \tag{6.79}$$

with T_F a constant. Then

$$[t^A, t^B] = if^{ABC} t^C; \qquad A, B, C = 1, \ldots, N^2 - 1, \tag{6.80}$$

where f^{ABC} is a set of constants (the *structure constants* of the group) completely antisymmetric in the three indices. We have inserted in Eq. (6.78) a coupling constant g in analogy with the phase transformations in electrodynamics. Let us assume that the Lagrangian density is invariant under the transformations (6.77), or, in infinitesimal form,

$$\delta\phi_i = ig\alpha^A t^A_{ij}\, \phi_j, \tag{6.81}$$

where summation on both indices A and j is understood. As we have seen in Sect. 2.2, the invariance of the Lagrangian density under the transformations (6.81) implies the existence of a set of conserved currents,

$$J^\mu_A = ig \frac{\partial \mathcal{L}}{\partial \partial_\mu \phi_i(x)}\, t^A_{ij}\, \phi_j(x). \tag{6.82}$$

The corresponding conserved charges do not commute.

Symmetry under local transformations, $\alpha^A = \alpha^A(x)$, is achieved by replacing ordinary derivatives by covariant derivatives, defined as

$$D^\mu = \partial^\mu I - igA^\mu, \tag{6.83}$$

where I is the unity matrix in the representation space, and the vector field A^μ is now a traceless hermitian matrix

$$A^\mu = A^\mu_A t^A. \tag{6.84}$$

It is easy to show, in analogy with the abelian case, that the transformation law

$$A^\mu \to A'^\mu = UA^\mu U^{-1} + \frac{i}{g} U\partial^\mu U^{-1} \tag{6.85}$$

ensures that

$$D^\mu \to U D^\mu U^{-1}. \tag{6.86}$$

To first order in the parameters α^A, Eq. (6.85) becomes

$$\begin{aligned} A'^\mu &= A^\mu + ig[\alpha^A t^A, A^\mu] - \frac{i}{g} ig\partial^\mu \alpha^A t^A \\ &= A^\mu_C t^C - g\alpha^A A^\mu_B f^{ABC} t^C + \partial^\mu \alpha^C t^C, \end{aligned} \tag{6.87}$$

or

$$A'^\mu_C = A^\mu_C - g\alpha^A A^\mu_B f^{ABC} + \partial^\mu \alpha^C. \tag{6.88}$$

A Lagrangian density for the vector fields A^μ_A can be built in analogy with the abelian case. Recalling Eq. (6.27), we define a traceless, hermitian matrix-valued field tensor $F^{\mu\nu}$ through

$$(D^\mu D^\nu - D^\nu D^\mu)\phi = -igF^{\mu\nu}\phi, \tag{6.89}$$

where ϕ is a multiplet of some $SU(N)$ representation, and $F^{\mu\nu} = F^{\mu\nu}_A t^A$. We find

$$\begin{aligned} F^{\mu\nu} &= \partial^\mu A^\nu - \partial^\nu A^\mu - ig[A^\mu, A^\nu], \\ F^{\mu\nu}_A &= \partial^\mu A^\nu_A - \partial^\nu A^\mu_A + gf^{ABC} A^\mu_B A^\nu_C. \end{aligned} \tag{6.90}$$

The Lagrangian density for the vector fields is then given by

$$\mathcal{L}_{\mathrm{YM}} = -\frac{1}{4} F^{\mu\nu}_A F^A_{\mu\nu}, \tag{6.91}$$

where the suffix YM stands for Yang and Mills. Since $F^{\mu\nu}_A$ contains terms quadratic in the vector fields, $\mathcal{L}_{\mathrm{YM}}$ contains self-interaction terms. This is related to the fact that, contrary to the abelian case, the field strength $F^{\mu\nu}$ transforms non-trivially under a gauge transformation:

$$F^{\mu\nu} \to F'^{\mu\nu} = UF^{\mu\nu}U^{-1}. \tag{6.92}$$

To first order in the parameters α^A,

$$F'^{\mu\nu}_A = F^{\mu\nu}_A - gf^{ABC}\alpha^B F^{\mu\nu}_C, \tag{6.93}$$

which means that the components $F^{\mu\nu}_A$ form a multiplet in the *adjoint representation* of the gauge group, whose generators are

$$T^A_{BC} = -if^{ABC}. \tag{6.94}$$

A simple calculation gives

$$\begin{aligned}\mathcal{L}_{\rm YM} = &-\frac{1}{4}\left(\partial^\mu A^\nu_A - \partial^\nu A^\mu_A\right)\left(\partial_\mu A_{\nu A} - \partial_\nu A_{\mu A}\right) \\ &-gf^{ABC}\partial_\mu A_{\nu A} A^\mu_B A^\nu_C \\ &-\frac{1}{4}g^2 f^{ABC} f^{ADE} A^\mu_B A^\nu_C A_{\mu D} A_{\nu E}.\end{aligned} \tag{6.95}$$

The Feynman rules are easily obtained; with a gauge-fixing term similar to the one we have chosen for electrodynamics, namely

$$\mathcal{L}_{\rm gf} = -\frac{1}{2}\left(\partial A^A\right)\left(\partial A^A\right), \tag{6.96}$$

we find the vector boson propagator

μ, A  ν, B $\leftrightarrow$ $\dfrac{g^{\mu\nu}}{q^2 + i\epsilon}\delta^{AB}.$ (6.97)

The interaction vertex with three vector bosons involves field derivatives. Following the usual procedure, we determine the corresponding vertex factor by evaluating the virtual amplitude that corresponds to the asymptotic field

$$A^{A\,({\rm as})}_\mu \sim e^{-ip_a\cdot x}, \tag{6.98}$$

which corresponds to assuming that momenta are incoming into the vertex. We obtain

ρ, C, p_c

μ, A, p_a 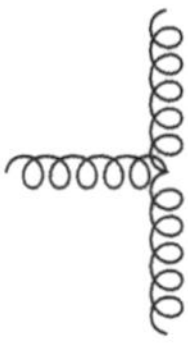 $igf^{ABC}\left[g^{\mu\rho}(p^\nu_a - p^\nu_c) + g^{\mu\nu}(p^\rho_b - p^\rho_a) + g^{\nu\rho}(p^\mu_c - p^\mu_b)\right].$

ν, B, p_b

(6.99)

Finally, the four vector boson vertex is given by:

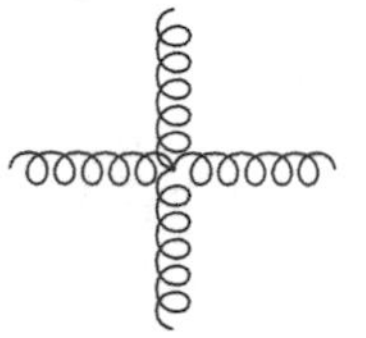

$$-g^2 f^{JAB} f^{JCD} \left(g^{\mu\rho} g^{\nu\sigma} - g^{\mu\sigma} g^{\nu\rho}\right)$$
$$-g^2 f^{JAC} f^{JBD} \left(g^{\mu\nu} g^{\rho\sigma} - g^{\mu\sigma} g^{\nu\rho}\right)$$
$$-g^2 f^{JAD} f^{JBC} \left(g^{\mu\nu} g^{\rho\sigma} - g^{\mu\rho} g^{\nu\sigma}\right) \qquad (6.100)$$

Strong interactions are now known to be described by a field theory that possesses a non-abelian gauge invariance under the group $SU(3)$. The associated non-commutative charges are called *colour charges*. This theory is obtained by coupling a pure $SU(3)$ Yang-Mills theory, whose vector bosons are usually called *gluons*, to a system of fermions, called *quarks*, that are assumed to transform according to the fundamental (dimension 3) representation of the invariance group. The Lagrangian density is built in analogy with electrodynamics:

$$\mathcal{L}_{\rm QCD} = \sum_f \bar{q}_f \left(i \not{D} - m_f\right) q_f + \mathcal{L}_{\rm YM} + \mathcal{L}_{\rm gf}, \qquad (6.101)$$

where the index f labels different quark species (usually called *flavours*), and the covariant derivative D is defined as

$$(D^\mu q_f)_i = [\partial^\mu \delta_i^j - i g_s A_A^\mu (t^A)_i^j] (q_f)_i, \qquad (6.102)$$

where i, j are colour indices. The generators t^A of $SU(3)$ in the fundamental representation are usually chosen as one half the familiar Gell-Mann matrices λ^A; in this case, $T_F = 1/2$. It is clear from Eq. (6.101) that strong interactions do not make distinctions among the different flavours. The nature of the flavour quantum number, and the origin of mass terms, is related to the fact that quarks also participate in the electroweak interactions, and will be described in Chaps. 7 and 9.

The above theory of strong interactions is usually called Quantum Chromodynamics (QCD), due to its close analogy with QED. The analogy extends to the invariance with respect to the T, C and P discrete transformations.

For completeness, we mention that non-abelian vector bosons have the same unphysical components as photons. However, here there is the further difficulty that, contrary to what happens in QED, $\partial_\mu A^\mu$ is not a free field. Thus the quantization of non-abelian gauge theories requires the introduction of a further set of unphysical fields, called *Faddeev-Popov ghosts*, with the role of compensating the effects of unphysical gluon components in the computation of probabilities. Further comments on the Faddeev-Popov ghosts can be found in Sect. 12.2.

Chapter 7
The Standard Model

7.1 A Gauge Theory of Weak Interactions

Weak interaction processes such as nucleon β decay, or μ decay, are correctly described by an effective theory, usually referred to as the Fermi theory of weak interactions. The Fermi Lagrangian density for β and μ decays is given by[1]

$$\mathcal{L} = -\frac{G^{(\beta)}}{\sqrt{2}}\overline{p}\gamma^{\alpha}(1 - a\gamma_5)n\,\overline{e}\gamma_{\alpha}(1 - \gamma_5)\nu_e - \frac{G^{(\mu)}}{\sqrt{2}}\overline{\nu}_{\mu}\gamma^{\alpha}(1 - \gamma_5)\mu\,\overline{e}\gamma_{\alpha}(1 - \gamma_5)\nu_e. \tag{7.1}$$

From the measured values of muon and neutron lifetimes,

$$\tau(N) = 885.7 \pm 0.8 \text{ s} \qquad \tau(\mu) = (2.19703 \pm 0.00004) \times 10^{-6} \text{ s}, \tag{7.2}$$

one obtains

$$G^{(\mu)} \simeq 1.16637(1) \times 10^{-5} \text{ GeV}^{-2}; \qquad G^{(\beta)} \simeq G^{(\mu)} \equiv G_F, \tag{7.3}$$

while the value

$$a = 1.2695 \pm 0.0029 \tag{7.4}$$

can be extracted from the measurement of baryon semi-leptonic decay rates. The most striking feature of weak interactions, correctly taken into account by the Fermi theory, is the violation of parity invariance, that arises from the measurement of neutrino helicities in weak decay processes. Note also that the Fermi constant sets a natural mass scale for weak interactions:

$$\Lambda \sim \frac{1}{\sqrt{G_F}} \sim 300 \text{ GeV}. \tag{7.5}$$

[1] Particle fields will be denoted by the symbols usually adopted for the corresponding particles: e for the electron, ν_e for the electron neutrino, and so on.

C. M. Becchi and G. Ridolfi, *An Introduction to Relativistic Processes and the Standard Model of Electroweak Interactions*, UNITEXT for Physics,
DOI: 10.1007/978-3-319-06130-6_7,

We note that the field theory defined by the interaction in Eq. (7.1) is non-renormalizable, since it contains operators with mass dimension 6; it is an effective theory, in the sense that it can only be used to compute amplitudes in the semi-classical approximation. Furthermore, we expect the results obtained by this effective theory to be accurate only when the energies of the particles involved are smaller than Λ. This condition is manifestly fulfilled by β and μ decays.

It can be shown (see Appendix G) that some $2 \to 2$ amplitudes computed with Eq. (7.1) grow quadratically with the total energy, and therefore eventually violate the unitarity bound. This confirms that the Fermi theory is only reliable at sufficiently low energies. This is a typical example of a general phenomenon. The presence of coefficients with negative mass dimension in the interaction induces, for dimensional reasons, the dominant terms in the amplitudes at large energies. This in turn originates a violation of the unitarity constraint Eq. (4.49). In the case of $2 \to 2$ particle amplitudes, if these grow as a positive power α of the energy E, the right-hand side of Eq. (4.49) grows as $E^{2\alpha}$, since the two-body phase space is asymptotically constant in energy, while the left-hand side grows at most as E^{α}.

Note that this source of violation of unitarity is not related to the semi-classical approximation: if one tries to recover unitarity by means of radiative corrections, as in the example of elastic scattering of scalar particles presented in Sect. 4.5, a series of terms is generated, that grow as a power of the energy which increases with the iterative order. At the same time, the divergence of Feynman diagram integrals grows order by order, and the theory is not renormalizable.

For these reasons, it is clear that a renormalizable and unitary theory of weak interactions is needed in order to perform precision calculations, to be compared with accurate measurements performed at large energies. We shall see that the Fermi theory contains the physical information needed to build such a theory.

The idea is that of building a theory with local invariance under the action of some group of transformations, a gauge theory, in analogy with quantum electrodynamics. The local four-fermion interaction of the Fermi Lagrangian will be interpreted as the interaction vertex that arises from the exchange of a massive vector boson with momentum much smaller than its mass. In this way, both problems of renormalizability and unitarity will be solved, since gauge theories are known to be renormalizable, and the mass of the intermediate vector boson will act as a cut-off that stops the growth of cross sections with energy, thus ensuring unitarity of the scattering matrix.

In order to complete this program, we must choose the group of local invariance, and then assign particle fields to representations of the chosen group. Both steps can be performed with the help of the information contained in the Fermi Lagrangian. Let us first consider the electron and the electron neutrino. They participate in the weak interaction via the current

$$J_\mu = \overline{\nu}_e \, \gamma_\mu \, \frac{1}{2}(1 - \gamma_5) \, e. \tag{7.6}$$

We would like to rewrite J_μ in the form of a Noether current,

$$\overline{\psi}_i \, \gamma_\mu \, T^A_{ij} \, \psi_j, \tag{7.7}$$

where ψ_i are the components of a multiplet in some representation of the (as yet unknown) gauge group, and T^A_{ij} are the corresponding generators. This can be done in the following way. We observe that the current J_μ can be written as

$$J_\mu = \overline{\ell}_L \gamma_\mu \tau^+ \ell_L, \tag{7.8}$$

where

$$\ell_L = \begin{pmatrix} \frac{1}{2}(1-\gamma_5)\,\nu_e \\ \frac{1}{2}(1-\gamma_5)\,e \end{pmatrix} \equiv \begin{pmatrix} \nu_{eL} \\ e_L \end{pmatrix}; \qquad \tau^+ = \frac{1}{2}(\tau_1 + i\tau_2) = \begin{pmatrix} 0 & 1 \\ 0 & 0 \end{pmatrix}, \tag{7.9}$$

and τ_i are the usual Pauli matrices,

$$\tau_1 = \begin{pmatrix} 0 & 1 \\ 1 & 0 \end{pmatrix}, \quad \tau_2 = \begin{pmatrix} 0 & -i \\ i & 0 \end{pmatrix}, \quad \tau_3 = \begin{pmatrix} 1 & 0 \\ 0 & -1 \end{pmatrix}. \tag{7.10}$$

The hermitian conjugate current

$$J^\dagger_\mu = \overline{\ell}_L \gamma_\mu \tau^- \ell_L; \qquad \tau^- = \frac{1}{2}(\tau_1 - i\tau_2) \tag{7.11}$$

will also participate in the interaction. As we have seen in Sect. 6.3, the currents are in one-to-one correspondence with the generators of the symmetry group, which, in turn, form a closed set with respect to the commutation operation. Therefore, the current

$$J^\mu_3 = \overline{\ell}_L \, \gamma^\mu \, [\tau^+, \tau^-] \, \ell_L = \overline{\ell}_L \, \gamma^\mu \, \tau_3 \, \ell_L \tag{7.12}$$

will also be present. No other current must be introduced, since

$$[\tau_3, \tau^\pm] = \pm 2\tau^\pm. \tag{7.13}$$

We have thus interpreted the current J_μ as one of the three conserved currents that arise from invariance under transformations of $SU(2)$ (it is known from ordinary quantum mechanics that the Pauli matrices are the generators of $SU(2)$ in the fundamental representation.) Furthermore, we have assigned the left-handed neutrino and electron fields to the fundamental representation of $SU(2)$, a *doublet*:

$$\begin{pmatrix} \nu_{eL} \\ e_L \end{pmatrix} \to \exp\left(ig\alpha_i \frac{\tau_i}{2}\right) \begin{pmatrix} \nu_{eL} \\ e_L \end{pmatrix}. \tag{7.14}$$

The right-handed neutrino and electron components, ν_{eR} and e_R, do not take part in the weak-interaction phenomena described by the Fermi Lagrangian, so they must be assigned to the singlet (or scalar) representation of $SU(2)$. For this reason, the invariance group $SU(2)$ relevant for weak interactions is sometimes referred to as $SU(2)_L$. Of course, this is not the only possible choice, but it is the simplest possibility since it does not require the introduction of fermion fields other than the observed ones.

The current J_3^μ is a *neutral* current: it is bilinear in creation and annihilation operators of particles with the same charge (actually, of the same particle.) Neutral currents do not appear in the Fermi Lagrangian, because no neutral current weak interaction phenomena were observed at the time of its formulation. As we shall see, the experimental observation of phenomena induced by weak neutral currents is a crucial test of the validity of the standard model. Notice also that the neutral current J_3^μ cannot be identified with the only other neutral current we know of, the electromagnetic current. There are two reasons for this: first, the electromagnetic current involves both left-handed and right-handed fermion fields with the same weight; and second, the electromagnetic current does not contain a neutrino term, the neutrino being chargeless. We will come back later to the problem of neutral currents, that will end up with the inclusion of the electromagnetic current in the theory. For the moment, we go on with the construction of our gauge theory based on $SU(2)$ invariance. We must introduce vector meson fields W_i^μ, one for each of the three $SU(2)$ generators, and build a covariant derivative

$$D^\mu = \partial^\mu - igW_i^\mu T_i, \tag{7.15}$$

where we have introduced a coupling constant g. The matrices T_i are generators of $SU(2)$ in the representation of the multiplet the covariant derivative is acting on. For example, when D^μ acts on the doublet ℓ_L, we have $T_i \equiv \tau_i/2$, and when it acts on the $SU(2)$ singlet e_R we have $T_i \equiv 0$. We are now ready to write the gauge-invariant Lagrangian for the fermion fields (which we assume massless for the time being):

$$\begin{aligned} \mathcal{L} &= i\overline{\ell}_L \not{D} \ell_L + i\overline{\nu}_{eR} \not{D} \nu_{eR} + i\overline{e}_R \not{D} e_R \\ &= \mathcal{L}_0 + \mathcal{L}_c + \mathcal{L}_n, \end{aligned} \tag{7.16}$$

where $\not{D} = \gamma_\mu D^\mu$. The Lagrangian $\mathcal{L}$ contains free terms for massless fermions,

$$\mathcal{L}_0 = i\overline{\ell}_L \not{\partial} \ell_L + i\overline{\nu}_{eR} \not{\partial} \nu_{eR} + i\overline{e}_R \not{\partial} e_R, \tag{7.17}$$

and an interaction term $\mathcal{L}_c + \mathcal{L}_n$, where

$$\mathcal{L}_c = gW_1^\mu \overline{\ell}_L \gamma_\mu \frac{\tau_1}{2} \ell_L + gW_2^\mu \overline{\ell}_L \gamma_\mu \frac{\tau_2}{2} \ell_L \tag{7.18}$$

corresponds to charged-current interactions, and

$$\mathcal{L}_n = gW_3^\mu \overline{\ell}_L \gamma_\mu \frac{\tau_3}{2} \ell_L = \frac{g}{2} W_3^\mu \left(\overline{\nu}_{eL} \gamma_\mu \nu_{eL} - \overline{e}_L \gamma_\mu e_L \right) \tag{7.19}$$

to neutral-current interactions. The charged-current term $\mathcal{L}_c$ is usually expressed in terms of the complex vector fields

$$W^{\pm}_{\mu} = \frac{1}{\sqrt{2}}(W^1_{\mu} \mp i W^2_{\mu}). \tag{7.20}$$

We find

$$\mathcal{L}_c = \frac{g}{\sqrt{2}}\bar{\ell}_L \gamma^{\mu} \tau^{+} \ell_L \, W^{+}_{\mu} + \frac{g}{\sqrt{2}}\bar{\ell}_L \gamma^{\mu} \tau^{-} \ell_L \, W^{-}_{\mu}. \tag{7.21}$$

We have already observed that the neutral current $J^{\mu}_3 = \bar{\ell}_L \gamma^{\mu} \tau_3 \ell_L$ cannot be identified with the electromagnetic current, and correspondingly that the gauge vector boson W^{μ}_3 cannot be interpreted as the photon field. The construction of the model can therefore proceed in two different directions: either we modify the multiplet structure of the theory, in order to make J^{μ}_3 equal to the electromagnetic current; or we admit the possibility of the existence of weak neutral currents, and we extend the gauge group in order to accommodate also the electromagnetic current in addition to J^{μ}_3. In the next Section we proceed to describe the second possibility, which turns out to be the correct one, after the discovery of weak processes induced by neutral currents. Nevertheless, it should be kept in mind that this was not at all obvious to physicists before the observation of weak neutral-current effects.

7.2 Electroweak Unification

The simplest way of extending the gauge group $SU(2)$ to include a second neutral generator is to include an abelian factor $U(1)$:

$$SU(2) \to SU(2) \times U(1). \tag{7.22}$$

We will require that the Lagrangian density be invariant also under the $U(1)$ gauge transformations

$$\psi \to \psi' = \exp\left(i g' \alpha \frac{Y(\psi)}{2}\right) \psi, \tag{7.23}$$

where ψ is a generic field in the theory, g' is the coupling constant associated to the $U(1)$ factor of the gauge group, and $Y(\psi)$ is a quantum number, usually called the *weak hypercharge*, to be specified for each field ψ with the condition that $Y(\psi)$ must be the same for all the components of the same $SU(2)$ multiplet since the charge Y must commute with all the $SU(2)$ generators.

Since the $SU(2)$ factor of the gauge group acts in a different way on left-handed and right-handed fermions (it is a *chiral* group), it is natural to allow for the possibility of assigning different hypercharge quantum numbers to left and right components of the same fermion field.

A new gauge vector field B^μ must be introduced, and the covariant derivative becomes

$$D^\mu = \partial^\mu - igW_i^\mu T_i - ig'\frac{Y}{2}B^\mu, \tag{7.24}$$

where Y is a diagonal matrix with the hypercharges in its diagonal entries. Since Y is diagonal, the introduction of the B^μ term in the covariant derivative only affects the neutral-current interaction $\mathcal{L}_n$. We have now

$$\begin{aligned}\mathcal{L}_n = &\frac{g}{2}W_3^\mu\left(\overline{\nu}_{eL}\gamma_\mu\nu_{eL} - \overline{e}_L\gamma_\mu e_L\right)\\ &+\frac{g'}{2}B^\mu\left[Y(\ell_L)\left(\overline{\nu}_{eL}\gamma_\mu\nu_{eL} + \overline{e}_L\gamma_\mu e_L\right) + Y(\nu_{eR})\overline{\nu}_{eR}\gamma_\mu\nu_{eR} + Y(e_R)\overline{e}_R\gamma_\mu e_R\right].\end{aligned} \tag{7.25}$$

This can be written as

$$\mathcal{L}_n = \overline{\Psi}\gamma_\mu\left[g\,T_3\,W_3^\mu + g'\,\frac{Y}{2}\,B^\mu\right]\Psi, \tag{7.26}$$

where Ψ is a column vector formed with all left-handed and right-handed fermion fields in the theory, and $T_3 = \pm 1/2$ for ν_{eL} and e_L respectively, and $T_3 = 0$ for ν_{eR} and e_R.

We can now assign the quantum numbers Y in such a way that the electromagnetic interaction term appears in Eq. (7.26). To do this, we perform a rotation by an angle θ_W in the space of the two neutral gauge fields W_3^μ, B^μ:

$$B^\mu = A^\mu \cos\theta_W - Z^\mu \sin\theta_W \tag{7.27}$$

$$W_3^\mu = A^\mu \sin\theta_W + Z^\mu \cos\theta_W\,. \tag{7.28}$$

In terms of the new vector fields A^μ, Z^μ, Eq. (7.26) takes the form

$$\begin{aligned}\mathcal{L}_n = &\overline{\Psi}\gamma_\mu\left(g\sin\theta_W T_3 + \frac{Y}{2}g'\cos\theta_W\right)\Psi\,A^\mu\\ &+\overline{\Psi}\gamma_\mu\left(g\cos\theta_W T_3 - \frac{Y}{2}g'\sin\theta_W\right)\Psi\,Z^\mu.\end{aligned} \tag{7.29}$$

In order to identify one of the two neutral vector fields, say A^μ, with the photon field, we must choose $Y(\ell_L)$, $Y(\nu_{eR})$ and $Y(e_R)$ so that A^μ couples to the electromagnetic current

$$J^\mu_{\text{em}} = -e\,\left(\overline{e}_R\gamma^\mu e_R + \overline{e}_L\gamma^\mu e_L\right) \equiv e\,\overline{\Psi}\,\gamma^\mu\,Q\,\Psi, \tag{7.30}$$

where Q is the diagonal matrix of electromagnetic charges in units of the positron charge e. In other words, we require that

$$T_3\, g \sin\theta_W + \frac{Y}{2}\, g' \cos\theta_W = e\, Q\,. \tag{7.31}$$

The weak hypercharges Y appear in Eq. (7.31) only through the combination Yg': thus, we have the freedom of rescaling the hypercharges by a common factor K, provided we rescale g' by $1/K$. This freedom can be used to fix arbitrarily the value of one of the three hypercharges $Y(\ell_L)$, $Y(\nu_{eR})$, $Y(e_R)$. The conventionally adopted choice is

$$Y(\ell_L) = -1. \tag{7.32}$$

With this choice, Eq. (7.31), restricted to the doublet of left-handed leptons, reads

$$+\frac{1}{2} g \sin\theta_W - \frac{1}{2} g' \cos\theta_W = 0 \tag{7.33}$$

$$-\frac{1}{2} g \sin\theta_W - \frac{1}{2} g' \cos\theta_W = -e, \tag{7.34}$$

which gives

$$g \sin\theta_W = g' \cos\theta_W = e. \tag{7.35}$$

(For a generic doublet of fields with charges Q_1 and Q_2, the r.h.s. of Eq. (7.35) becomes $e(Q_1 - Q_2)$, but charge conservation requires $Q_1 - Q_2 = 1$.) Equation (7.31) then reduces to

$$T_3 + \frac{Y}{2} = Q. \tag{7.36}$$

Therefore, we find

$$Y(\nu_{eR}) = 0; \qquad Y(e_R) = -2. \tag{7.37}$$

This completes the assignments of weak hypercharges to all fermion fields. Note that the right-handed neutrino is an $SU(2)$ singlet with zero hypercharge: it does not take part in electroweak interactions, and can be omitted altogether. We shall consider the possible effects of right-handed neutrinos in Chap. 13.

The second term in Eq. (7.29) defines the weak neutral current coupled to the weak neutral vector boson Z_μ. It can be written as

$$e\,\overline{\Psi}\gamma_\mu\, Q_Z\, \Psi\, Z^\mu\,, \tag{7.38}$$

where

$$Q_Z = \frac{1}{\cos\theta_W \sin\theta_W}\left(T_3 - Q\sin^2\theta_W\right). \tag{7.39}$$

This structure can be replicated for an arbitrary numbers of lepton families, assigned to the same representations of $SU(2) \times U(1)$ as the electron and the electron neutrino. At present, the existence of two more lepton families is experimentally established: the lepton μ^- with its neutrino ν_μ, and lepton τ^- with its neutrino ν_τ.

7.3 Hadrons

We must now include hadrons into the theory. We will do this in terms of quark fields, taking as a starting point the hadronic current responsible for β decay and strange particle decays:

$$J^{\mu}_{\text{hadr}} = \cos\theta_c\,\overline{u}\gamma^{\mu}\frac{1}{2}(1-\gamma_5)d + \sin\theta_c\,\overline{u}\gamma^{\mu}\frac{1}{2}(1-\gamma_5)s, \tag{7.40}$$

where θ_c is the Cabibbo angle ($\theta_c \sim 13°$) and u, d, s are the up, down and strange quark fields respectively (for simplicity, we have not displayed the colour indices carried by quarks.) As noted in Sect. 6.3, different quark flavours are not distinguished by strong interactions. The corresponding particles have however different masses, which are of the Dirac type because of various symmetry reasons. In Eq. (7.40), the fields u, d, s represent definite mass eigenfields. Lepton and quark masses will be discussed in detail in Chap. 9; for the time being, we just assumed that quarks with different flavours can be distinguished from each other.

We are tempted to proceed as in the case of leptons: we define

$$q_L = \frac{1}{2}(1-\gamma_5)\begin{bmatrix} u \\ d \\ s \end{bmatrix} \equiv \begin{bmatrix} u_L \\ d_L \\ s_L \end{bmatrix} \tag{7.41}$$

and

$$T^{+} = \begin{bmatrix} 0 & \cos\theta_c & \sin\theta_c \\ 0 & 0 & 0 \\ 0 & 0 & 0 \end{bmatrix}, \tag{7.42}$$

so that

$$J^{\mu}_{\text{hadr}} = \overline{q}_L\,\gamma^{\mu}\,T^{+}\,q_L. \tag{7.43}$$

This procedure would lead to a system of currents which is in contrast with experimental observations. Indeed, we find

$$T_3 = [T^{+}, T^{-}] = \begin{bmatrix} 1 & 0 & 0 \\ 0 & -\cos^2\theta_c & -\cos\theta_c\sin\theta_c \\ 0 & -\cos\theta_c\sin\theta_c & -\sin^2\theta_c \end{bmatrix}. \tag{7.44}$$

The corresponding neutral current contains flavour-changing terms, such as for example $\overline{d}_L\gamma^{\mu}s_L$, with a weight of the same order of magnitude of flavour-conserving ones. These terms induce processes at a rate which is not compatible with data. For example, the ratio of the decay rates for the processes

$$K^{+} \to \pi^{0}e^{+}\nu_e \tag{7.45}$$

$$K^{+} \to \pi^{+}e^{+}e^{-} \tag{7.46}$$

would be approximately

$$r = \left[\frac{\sin\theta_c}{\sin\theta_c \cos\theta_c} \right]^2 = \frac{1}{\cos^2\theta_c} \simeq 1.1, \tag{7.47}$$

while observations give

$$r_{\exp} \simeq 1.3 \times 10^5, \tag{7.48}$$

that is, the charged-current process ($s \to u$) is enhanced by five orders of magnitude with respect to the neutral-current process ($s \to d$). Our theory should therefore be modified in order to avoid the introduction of flavour-changing neutral currents. The solution to this puzzle was found by S. Glashow, J. Iliopoulos and L. Maiani. They suggested to introduce a fourth quark c (for *charm*) with charge 2/3 like the up quark, and to assume that its couplings to down and strange quarks are given by

$$\begin{aligned} J^\mu_{\text{hadr}} &= \cos\theta_c\, \overline{u}\gamma^\mu \frac{1}{2}(1-\gamma_5)d + \sin\theta_c\, \overline{u}\gamma^\mu \frac{1}{2}(1-\gamma_5)s \\ &\quad - \sin\theta_c\, \overline{c}\gamma^\mu \frac{1}{2}(1-\gamma_5)d + \cos\theta_c\, \overline{c}\gamma^\mu \frac{1}{2}(1-\gamma_5)s. \end{aligned} \tag{7.49}$$

The c quark being not observed at the time, they had to assume that its mass was much larger than those of u, d and s quarks, and therefore outside the energy range of available experimental devices. The current J^μ_{hadr} can still be cast in the form of Eq. (7.43), where now

$$q_L = \begin{bmatrix} u_L \\ c_L \\ d_L \\ s_L \end{bmatrix}; \qquad T^+ = \begin{bmatrix} 0 & 0 & \cos\theta_c & \sin\theta_c \\ 0 & 0 & -\sin\theta_c & \cos\theta_c \\ 0 & 0 & 0 & 0 \\ 0 & 0 & 0 & 0 \end{bmatrix}. \tag{7.50}$$

No flavour-changing neutral current is now present, since

$$[T^+, T^-] = \begin{bmatrix} 1 & 0 & 0 & 0 \\ 0 & 1 & 0 & 0 \\ 0 & 0 & -1 & 0 \\ 0 & 0 & 0 & -1 \end{bmatrix}, \tag{7.51}$$

as a consequence of the fact that the upper right 2×2 block of T^+ was cleverly chosen to be an orthogonal matrix. The existence of the quark c was later confirmed by the discovery of the J/ψ particle. The current J^μ_{hadr} is usually written in the following form, analogous to the corresponding leptonic current:

$$J^\mu_{\text{hadr}} = (\overline{u}_L \overline{d}'_L)\gamma^\mu \tau^+ \begin{pmatrix} u_L \\ d'_L \end{pmatrix} + (\overline{c}_L \overline{s}'_L)\gamma^\mu \tau^+ \begin{pmatrix} c_L \\ s'_L \end{pmatrix}, \tag{7.52}$$

where

$$\begin{pmatrix} d'_L \\ s'_L \end{pmatrix} = V \begin{pmatrix} d_L \\ s_L \end{pmatrix}, \qquad V = \begin{pmatrix} \cos\theta_c & \sin\theta_c \\ -\sin\theta_c & \cos\theta_c \end{pmatrix}. \tag{7.53}$$

The pairs (u, d), (c, s) are called *quark families*. The structure outlined above can be extended to an arbitrary number of quark families. With n families, V becomes an $n \times n$ matrix, and it must be unitary in order to ensure the absence of flavour-changing neutral currents. One more quark family, in addition to (u, d) and (c, s), has been detected, namely the one formed by the *top* and the *bottom* quarks (t, b). In the following, quark families will be denoted by (u^f, d^f), with the index f running over the families. We will show in Chap. 12 that the number of lepton and quark families must be equal for the theory to be consistent.

The final form of the standard model Lagrangian with n families of leptons and quarks is then

$$\mathcal{L} = \mathcal{L}_0 + \mathcal{L}_c + \mathcal{L}_n \tag{7.54}$$

$$\begin{aligned} \mathcal{L}_0 = {} & i\bar{\ell}^f_L \not{\partial} \ell^f_L + i\bar{\nu}^f_R \not{\partial} \nu^f{}_R + i\bar{e}^f_R \not{\partial} e^f_R \\ & + i\bar{q}^f_L \not{\partial} q^f_L + i\bar{u}^f{}_R \not{\partial} u^f{}_R + i\bar{d}^f_R \not{\partial} d^f_R \end{aligned} \tag{7.55}$$

$$\mathcal{L}_n = e\overline{\Psi}\gamma_\mu Q \Psi A^\mu + e\overline{\Psi}\gamma_\mu Q_Z \Psi Z^\mu \tag{7.56}$$

$$\mathcal{L}_c = \frac{g}{\sqrt{2}} \sum_{f=1}^{n} \left(\bar{\ell}^f_L \gamma^\mu \tau^+ \ell^f_L + \bar{q}^f_L \gamma^\mu \tau^+ q^f_L \right) W^+_\mu + \text{h.c.}, \tag{7.57}$$

where

$$\ell^f_L = \begin{pmatrix} \nu_{eL} \\ e_L \end{pmatrix}, \begin{pmatrix} \nu_{\mu L} \\ \mu_L \end{pmatrix}, \begin{pmatrix} \nu_{\tau L} \\ \tau_L \end{pmatrix}, \ldots \tag{7.58}$$

$$q^f_L = \begin{pmatrix} u_L \\ d'_L \end{pmatrix}, \begin{pmatrix} c_L \\ s'_L \end{pmatrix}, \begin{pmatrix} t_L \\ b'_L \end{pmatrix}, \ldots. \tag{7.59}$$

The vector Ψ now also includes left-handed and right-handed quark fields. Right-handed quark fields are assigned to $SU(2)$ singlets, with hypercharges given by Eq. (7.36) (we recall that $Q = 2/3$ for $u, c, t, \ldots$ and $Q = -1/3$ for $d, s, b, \ldots$.) and Q_Z is given by the same expression Eq. (7.39), with $T_3 = +1/2$ for u_L, $T_3 = -1/2$ for d_L, $T_3 = 0$ for right-handed quarks.

An equivalent (and often more useful) form of Eq. (7.57) is

$$\mathcal{L}_c = \frac{g}{\sqrt{2}} \left(\sum_{f=1}^{n} \bar{\nu}^f_L \gamma^\mu e^{f_L} + \sum_{f,g=1}^{n} \bar{u}^{f_L} \gamma^\mu V_{fg} d^g_L \right) W^+_\mu + \text{h.c.} \tag{7.60}$$

To conclude the construction of the standard model, we must include the vector boson Lagrangian density

$$\mathcal{L}_{\mathrm{YM}} = -\frac{1}{4} B_{\mu\nu} B^{\mu\nu} - \frac{1}{4} W^i_{\mu\nu} W_i^{\mu\nu}, \tag{7.61}$$

where

$$\begin{aligned} B^{\mu\nu} &= \partial^\mu B^\nu - \partial^\nu B^\mu \\ W_i^{\mu\nu} &= \partial^\mu W_i^\nu - \partial^\nu W_i^\mu + g\epsilon_{ijk} W_j^\mu W_k^\nu . \end{aligned} \tag{7.62}$$

The corresponding expressions in terms of $W_\mu^\pm$, Z_μ and A_μ can be easily worked out with the help of Eqs. (7.20), (7.27) and (7.28), which we repeat here:

$$W^1_\mu = \frac{1}{\sqrt{2}}(W_\mu^+ + W_\mu^-) \tag{7.63}$$

$$W^2_\mu = \frac{i}{\sqrt{2}}(W_\mu^+ - W_\mu^-) \tag{7.64}$$

$$W^3_\mu = A_\mu \sin\theta_W + Z_\mu \cos\theta_W \tag{7.65}$$

$$B_\mu = A_\mu \cos\theta_W - Z_\mu \sin\theta_W . \tag{7.66}$$

We get

$$\begin{aligned} W^1_{\mu\nu} &= \frac{1}{\sqrt{2}}\Big[W^+_{\mu\nu} + ig \sin\theta_W (W_\mu^+ A_\nu - W_\nu^+ A_\mu) \\ &\qquad + ig\cos\theta_W (W_\mu^+ Z_\nu - W_\nu^+ Z_\mu)\Big] + \text{h.c.} \\ W^2_{\mu\nu} &= \frac{i}{\sqrt{2}}\Big[W^+_{\mu\nu} + ig \sin\theta_W (W_\mu^+ A_\nu - W_\nu^+ A_\mu) \\ &\qquad + ig\cos\theta_W (W_\mu^+ Z_\nu - W_\nu^+ Z_\mu)\Big] + \text{h.c.} \\ W^3_{\mu\nu} &= F_{\mu\nu}\sin\theta_W + Z_{\mu\nu}\cos\theta_W - ig(W_\mu^+ W_\nu^- - W_\mu^- W_\nu^+) \\ B_{\mu\nu} &= F_{\mu\nu}\cos\theta_W - Z_{\mu\nu}\sin\theta_W , \end{aligned} \tag{7.67}$$

where

$$F^{\mu\nu} = \partial^\mu A^\nu - \partial^\nu A^\mu \tag{7.68}$$

$$Z^{\mu\nu} = \partial^\mu Z^\nu - \partial^\nu Z^\mu \tag{7.69}$$

$$W_\pm^{\mu\nu} = \partial^\mu W_\pm^\nu - \partial^\nu W_\pm^\mu . \tag{7.70}$$

It follows that

$$
\begin{aligned}
\mathcal{L}_{\mathrm{YM}} =& -\frac{1}{4}F_{\mu\nu}F^{\mu\nu} - \frac{1}{4}Z_{\mu\nu}Z^{\mu\nu} - \frac{1}{2}W^{+}_{\mu\nu}W^{\mu\nu}_{-} \\
&+ ig\sin\theta_W\,(W^{+}_{\mu\nu}W^{\mu}_{-}A^{\nu} - W^{-}_{\mu\nu}W^{\mu}_{+}A^{\nu} + F_{\mu\nu}W^{\mu}_{+}W^{\nu}_{-}) \\
&+ ig\cos\theta_W\,(W^{+}_{\mu\nu}W^{\mu}_{-}Z^{\nu} - W^{-}_{\mu\nu}W^{\mu}_{+}Z^{\nu} + Z_{\mu\nu}W^{\mu}_{+}W^{\nu}_{-}) \\
&+ \frac{g^2}{2}(2g^{\mu\nu}g^{\rho\sigma} - g^{\mu\rho}g^{\nu\sigma} - g^{\mu\sigma}g^{\nu\rho})\Big[\frac{1}{2}W^{+}_{\mu}W^{+}_{\nu}W^{-}_{\rho}W^{-}_{\sigma} \\
&- W^{+}_{\mu}W^{-}_{\nu}(A_\rho A_\sigma \sin^2\theta_W + Z_\rho Z_\sigma \cos^2\theta_W + 2A_\rho Z_\sigma \sin\theta_W\cos\theta_W)\Big].
\end{aligned}
\tag{7.71}
$$

Chapter 8
Spontaneous Breaking of the Gauge Symmetry

8.1 Masses for Vector Bosons

We have seen in Chap. 7 that the neutral vector boson fields coupled to the $SU(2)$ and $U(1)$ charges appear as linear combinations of the massless photon field and of the Z^0 field; similarly, left-handed quarks with charge $-1/3$ enter the weak interaction term as linear combinations of the corresponding mass eigenfields. We show here that these mixing phenomena are consequences of the mechanism of mass generation in the model.

We first show that, in order to make contact with the Fermi theory,which is known to describe correctly weak interactions at low energies, the gauge vector bosons of weak interactions must have non-vanishing masses. On the same basis, we will also be able to set a lower bound to the mass of the W boson. Let us consider the amplitude for down-quark β decay. In the Fermi theory, it is simply given by

$$\mathcal{M} = -\frac{G_F}{\sqrt{2}}\overline{u}\gamma^\mu(1-\gamma_5)d\,\overline{e}\gamma_\mu(1-\gamma_5)\nu_e. \tag{8.1}$$

In the context of the standard model, the same process is induced by the exchange of a W boson, with amplitude

$$\mathcal{M}^{\text{SM}} = \left(\frac{g}{\sqrt{2}}\overline{u}_L\gamma^\mu d_L\right)\frac{1}{q^2-m_W^2}\left(\frac{g}{\sqrt{2}}\overline{e}_L\gamma_\mu\nu_{eL}\right), \tag{8.2}$$

(we are neglecting the Cabibbo angle for simplicity). The virtuality q^2 of the exchanged vector boson is bounded from above by the square of the neutron-proton mass difference, $q^2 \le (m_{\text{N}}-m_{\text{P}})^2 \sim (1.3\ \text{MeV})^2$. Equation (8.2) reduces to the Fermi amplitude provided $m_W^2 \gg q^2$, and

$$\frac{G_F}{\sqrt{2}} = \left(\frac{g}{2\sqrt{2}}\right)^2\frac{1}{m_W^2}. \tag{8.3}$$

C. M. Becchi and G. Ridolfi, *An Introduction to Relativistic Processes and the Standard Model of Electroweak Interactions*, UNITEXT for Physics,
DOI: 10.1007/978-3-319-06130-6_8, © Springer International Publishing Switzerland 2014

Recalling that $g = e/\sin\theta_W$, Eq. (8.3) gives the lower bound

$$m_W \geq 37.3 \text{ GeV}, \tag{8.4}$$

which is quite a large value, compared to the nucleon mass, and an enormous number, compared to the present upper bound on the photon mass,

$$m_\gamma \leq 2 \cdot 10^{-16} \text{ eV}. \tag{8.5}$$

We conclude that if weak interactions are to be mediated by vector bosons, these must be very heavy. On the other hand, we also know that gauge theories are incompatible with mass terms for the vector bosons.

One possible way out is breaking gauge invariance explicitly; this, however, as already observed in Sect. 6.1, leads to a non-renormalizable and non-unitary theory. Let us investigate this point in more detail. One may formulate a theory for a massive gauge boson with the propagator

$$\Delta^{\mu\nu}(k) = \frac{1}{k^2 - \mu^2}\left(g^{\mu\nu} - \frac{k^\mu k^\nu}{\mu^2}\right), \tag{8.6}$$

so that non-physical components of the vector field do not contribute. Notice however that weak charged current conservation is necessary in order to reproduce the Fermi interaction in Eq. (8.1). Indeed otherwise one should have a further term associated with the second term of the propagator in Eq. (8.6). This terms would involve scalar, rather than vector, densities, which have been excluded by the experimental analyses of weak decays. Furthermore, for large values of the momentum k, the term proportional to $k^\mu k^\nu$ in the propagator Eq. (8.6) dominates; it is therefore clear that the behaviour of this propagator at large k is much worse than that of the scalar propagator, that vanishes at infinity as $1/k^2$. This suggests that the propagator Eq. (8.6) leads to a non-renormalizable theory.

A related problem of a massive vector boson theory without gauge invariance is unitarity of the scattering matrix. The amplitude for a generic physical process which involves the emission or the absorption of a vector boson with four-momentum k and polarization vector $\epsilon(k)$ has the form

$$\mathcal{M} = \mathcal{M}^\mu \epsilon_\mu(k). \tag{8.7}$$

A massive vector (contrary to a massless one) may be polarized longitudinally. In this case, choosing the z axis along the direction of the 3-momentum of the vector boson, the polarization is given by

$$\epsilon_L = \left(\frac{|\boldsymbol{k}|}{\mu}, 0, 0, \frac{E}{\mu}\right) = \frac{k}{\mu} + \mathcal{O}(\mu^2/E^2), \tag{8.8}$$

where we have imposed the transversity condition $k \cdot \epsilon = 0$ and the normalization condition $\epsilon^2 = -1$. Clearly, the amplitude $\mathcal{M}$ will grow indefinitely with the energy E, thus eventually violating the unitarity bound.

Both sources of power-counting violation are rendered harmless if the vector particles are coupled to conserved currents, thus confirming the need of gauge invariance.

8.2 Scalar Electrodynamics and the Abelian Higgs Model

In order to see how one can introduce a mass term for a gauge vector boson without spoiling renormalizability and unitarity, we first consider a simple example, and then we generalize our considerations to the standard model. The simple theory we consider is scalar electrodynamics, that is, a $U(1)$ gauge theory for one complex scalar field Φ with charge e. The requirement of invariance under the gauge transformations

$$\Phi(x) \to e^{ie\Lambda(x)}\Phi(x) \tag{8.9}$$
$$\Phi^*(x) \to e^{-ie\Lambda(x)}\Phi^*(x) \tag{8.10}$$
$$A_\mu(x) \to A_\mu(x) + \partial_\mu \Lambda(x) \tag{8.11}$$

and the gauge choice Eq. (6.38) lead uniquely to the Lagrangian density

$$\mathcal{L} = D_\mu \Phi^* \, D^\mu \Phi - m^2 \Phi^* \Phi - \lambda \left(\Phi^* \Phi\right)^2 - \frac{1}{2}\partial_\mu A_\nu \partial^\mu A^\nu, \tag{8.12}$$

where

$$D_\mu \Phi(x) = \partial_\mu \Phi(x) - ieA_\mu(x)\Phi(x). \tag{8.13}$$

Observe that scalar QED involves one more coupling constant (λ) than spin-1/2 QED.

The field equations are given by

$$\left(\partial^2 + m^2\right)\Phi = ieA^\mu \partial_\mu \Phi + ie\partial_\mu \left(A^\mu \Phi\right) + e^2 A^2 \Phi - 2\lambda \Phi^* \Phi^2 \equiv J_\Phi \tag{8.14}$$
$$\partial^2 A^\mu = -ie\left(\Phi^* D^\mu \Phi - \Phi D^\mu \Phi^*\right) \equiv -J_A^\mu. \tag{8.15}$$

The Green function for the scalar field is $\Delta(x)$, Eq. (3.100), while the Green function for the vector field A_μ is the same as in Sect. 6.1.

The gauge transformations Eqs. (8.9, 8.10) have the property that they leave the origin of the isotopic space unchanged. There is however the possibility of defining gauge transformations on scalar fields that do not have this feature; let us explore this possibility. We require invariance under the transformations

$$\Phi(x) \to e^{ie\Lambda(x)}\left(\Phi(x)+\frac{v}{\sqrt{2}}\right)-\frac{v}{\sqrt{2}} \tag{8.16}$$

$$\Phi^*(x) \to e^{-ie\Lambda(x)}\left(\Phi^*(x)+\frac{v}{\sqrt{2}}\right)-\frac{v}{\sqrt{2}} \tag{8.17}$$

$$A_\mu(x) \to A_\mu(x)+\partial_\mu\Lambda(x), \tag{8.18}$$

where v is a real constant. The covariant derivative term must be consistently modified. We observe that the quantity

$$D_\mu\left(\Phi(x)+\frac{v}{\sqrt{2}}\right) \equiv \partial_\mu\Phi(x) - ieA_\mu(x)\left(\Phi(x)+\frac{v}{\sqrt{2}}\right) \tag{8.19}$$

has the same transformation property as $\Phi + \frac{v}{\sqrt{2}}$; thus, it can be taken as the definition of the covariant derivative when the gauge transformation is defined as in Eqs. (8.16–8.18). The self-interaction of the field Φ is entirely determined by the requirements of renormalizability and invariance under the gauge transformations Eqs. (8.16–8.18). The further condition, discussed in Sect. 2.1, that the scalar potential have a minimum at the origin of the isotopic space, fixes the form of the scalar potential up to an overall constant:

$$V(\Phi) = \lambda\left[\left(\Phi(x)+\frac{v}{\sqrt{2}}\right)\left(\Phi^*(x)+\frac{v}{\sqrt{2}}\right)-\frac{v^2}{2}\right]^2. \tag{8.20}$$

In this context, it is convenient to replace the gauge-fixing term $\mathcal{L}_{gf}$ of Eq. (6.38) with

$$\mathcal{L}_{gfh} = -\frac{1}{2}\left(\partial_\mu A^\mu - ie\frac{v}{\sqrt{2}}(\Phi-\Phi^*)\right)^2, \tag{8.21}$$

first introduced by G.'t Hooft. We recall that the gauge-fixing term provides a gauge condition as a consequence of the field equations; the choice of one particular gauge condition is not related to physical considerations. The gauge-fixing term completes the formulation of a generalization of scalar QED, which is called the *abelian Higgs model*. The Lagrangian density is

$$\begin{aligned}\mathcal{L}_H =& \left[D_\mu\left(\Phi+\frac{v}{\sqrt{2}}\right)\right]^* D^\mu\left(\Phi+\frac{v}{\sqrt{2}}\right) - V(\Phi)\\ & -\frac{1}{4}F_{\mu\nu}F^{\mu\nu} - \frac{1}{2}\left(\partial_\mu A^\mu - ie\frac{v}{\sqrt{2}}(\Phi-\Phi^*)\right)^2.\end{aligned} \tag{8.22}$$

The particle content of the theory is immediately determined by inspection of the free Lagrangian density

$$
\begin{aligned}
\mathcal{L}_0 =& \partial_\mu \Phi^* \partial^\mu \Phi + ie\frac{v}{\sqrt{2}} A^\mu \partial_\mu (\Phi - \Phi^*) + \frac{e^2 v^2}{2} A^2 - \frac{\lambda}{2} v^2 (\Phi + \Phi^*)^2 \\
&- \frac{1}{2} \partial_\mu A_\nu \partial^\mu A^\nu + \frac{1}{2} \partial_\mu A_\nu \partial^\nu A^\mu - \frac{1}{2} (\partial A)^2 + \frac{e^2 v^2}{4} (\Phi - \Phi^*)^2 \\
&+ ie\frac{v}{\sqrt{2}} \partial_\mu A^\mu (\Phi - \Phi^*) \\
=& \partial_\mu \Phi^* \partial^\mu \Phi - \frac{\lambda}{2} v^2 (\Phi + \Phi^*)^2 + \frac{e^2 v^2}{4} (\Phi - \Phi^*)^2 - \frac{1}{2} \partial_\mu A_\nu \partial^\mu A^\nu + \frac{e^2 v^2}{2} A^2.
\end{aligned}
\tag{8.23}
$$

In the second step we have performed a partial integration on the last term; we see that the gauge-fixing term was chosen so that bilinear terms proportional to $A^\mu \partial_\mu (\Phi - \Phi^*)$ cancel. We have also omitted a term proportional to a four-divergence, in analogy with ordinary electrodynamics (see Eq. (6.40)). It is convenient to decompose the field Φ into its real and imaginary parts:

$$
\Phi(x) = \frac{H(x) + iG(x)}{\sqrt{2}}. \tag{8.24}
$$

The real scalar field H is usually called the *Higgs field*, while G is a *Goldstone field*. We find

$$
\mathcal{L}_0 = \frac{1}{2} \left[(\partial H)^2 - 2\lambda v^2 H^2 + (\partial G)^2 - e^2 v^2 G^2 - \partial_\mu A_\nu \partial^\mu A^\nu + e^2 v^2 A^2 \right]. \tag{8.25}
$$

We see that the model contains spin 1 and spin 0 bosons with mass ev, associated with the field A, and two different kinds of scalars: the Goldstone boson, with the same mass, and the Higgs boson, with mass $\sqrt{2\lambda} v$.

The phenomenon we have just described is usually called *Higgs mechanism for the spontaneous breaking of the gauge symmetry*, even though the symmetry is not actually broken. In fact, the Lagrangian density is still gauge invariant, up to the gauge-fixing term, which however does not influence physical quantities. Hence, all physical properties connected with gauge invariance (such as, for example, current conservation within the space of physical states) are still there. It is important to stress this point, because at the quantum level this is essentially what guarantees the renormalizability of the theory, which would instead be lost in the case of an explicit breaking of the gauge symmetry.

The interaction term is given by

$$
\begin{aligned}
\mathcal{L}_I =& - eA^\mu \left(H \partial_\mu G - G \partial_\mu H \right) + e^2 v A^2 H - \lambda v H \left(H^2 + G^2 \right) \\
&+ \frac{e^2}{2} A^2 H^2 + \frac{e^2}{2} A^2 G^2 - \frac{\lambda}{4} \left(H^4 + G^4 + 2H^2 G^2 \right).
\end{aligned}
\tag{8.26}
$$

The field equations for A^μ, H and G can be written in full analogy with Eqs. (3.90, 8.15):

$$\left(\partial^2 + 2\lambda v^2\right) H = J_H \tag{8.27}$$

$$\left(\partial^2 + e^2 v^2\right) G = J_G \tag{8.28}$$

$$\left(\partial^2 + e^2 v^2\right) A^\mu = -J^\mu, \tag{8.29}$$

where the r.h.s. can be directly computed from the interaction Lagrangian.

The Feynman rules for the abelian Higgs model can be worked out by the same procedure adopted in the case of the scalar theory. The propagators for scalar fields are obtained from Eq. (3.100) replacing the appropriate values of the masses; that of the vector field is obtained from Eq. (6.44) inserting the mass squared e^2v^2 in the denominator. The correspondence between line factors and symbols is therefore:

 $\leftrightarrow \quad \dfrac{1}{2\lambda v^2 - q^2 - i\epsilon}$ for the field H (8.30)

 $\leftrightarrow \quad \dfrac{1}{e^2 v^2 - q^2 - i\epsilon}$ for the field G (8.31)

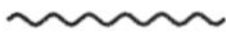

 $\leftrightarrow \quad \dfrac{g^{\mu\nu}}{q^2 - e^2 v^2 + i\epsilon}$ for the field A^μ. (8.32)

Using the same symbols for lines, the interaction vertices induced by the interaction terms Eq. (8.26) are given by

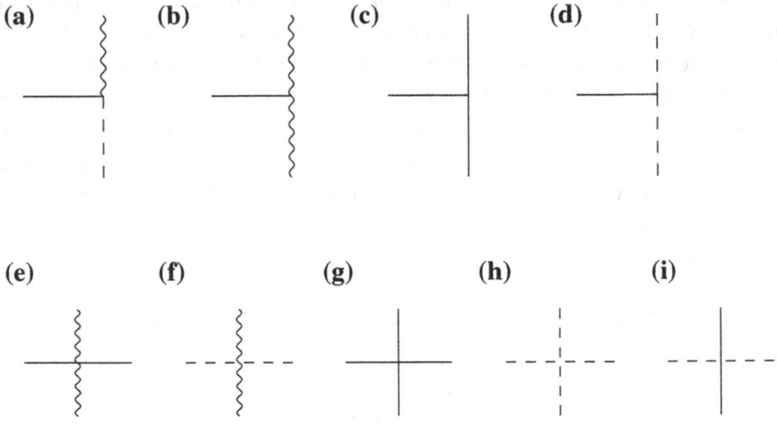

Table 8.1 Feynman rules for the interaction vertices in the abelian Higgs model

(a)	(b)	(c)	(d)	(e)	(f)	(g)	(h)	(j)
$ie\left(p_G^\mu - p_H^\mu\right)$	$2e^2 v g^{\mu\nu}$	$-6\lambda v$	$-2\lambda v$	$2e^2 g^{\mu\nu}$	$2e^2 g^{\mu\nu}$	-6λ	-6λ	-2λ

The vertex (a) is originated by the first term in Eq. (8.26); it corresponds to an interaction term that involves derivative of the fields. Following the usual procedure, we obtain

$$ie\left(p_G^\mu - p_H^\mu\right), \tag{8.33}$$

where the four-momenta p_H of H and p_G of G are taken to be incoming in the vertex. All other interaction vertex factors are simply given by the corresponding coefficient in the interaction Lagrangian, multiplied with the combinatorial factor $\prod_c n_c!$, where n_c is the power of the field of species c in the vertex. The vertex factors are listed in Table 8.1.

Notice that in the limit $e \to 0$, $\lambda \to 0$ and $v \to \infty$ with $\lambda v^2 \equiv M^2$ and $ev \equiv m$ fixed the abelian Higgs model becomes a free theory involving two scalar and one vector field, all of them massive. The vector field can now be coupled to some other matter fields, such as e.g. fermions as in the case of spinor QED, through a charge $\hat{e}$ which is kept fixed in the above limit; the resulting theory is a massive version of QED, which was originally discovered by Stueckelberg. Both scalar fields remain free and can be omitted. This shows that, in practice, one can simply add a mass term to the Lagrangian of QED, without actually spoiling the internal consistency of the theory.

However, the same procedure cannot be applied to non-abelian gauge theories (based e.g. on $SU(N)$ invariance.) In that case, the gauge fields are coupled to matter fields through a coupling constant g, which is the same that appears in self-interaction terms. Therefore, non-abelian gauge-invariance becomes abelian in the limit $g \to 0$.

8.3 Physical Content of the Abelian Higgs Model

It is interesting to study the physical content of the abelian Higgs model described in the previous section, since many of its features are shared by its extension to cases of practical interest, to be discussed later in this chapter.

The spin of the particles involved is related to the transformation properties of the corresponding fields under spatial rotations: scalars correspond to spin 0, three-vectors to spin 1. Thus, the Higgs and Goldstone fields correspond to spinless particles. The massive four-vector field contains four degrees of freedom, three of which correspond to a spin-1 particle, and one to spin 0. To see this, we observe that from the free-field Lagrangian Eq. (8.25) one obtains an expression of the free field in terms of creation and annihilation operators, that allows quantization by analogy with the simple harmonic oscillator:

$$A_\mu(x) = \frac{1}{(2\pi)^{3/2}} \int \frac{d^3k}{\sqrt{2E_k}} \sum_{h=0}^{3} \left[a^{(h)}(\boldsymbol{k})\epsilon_\mu^{(h)}(\boldsymbol{k})\, e^{-ikx} + a^{\dagger(h)}(\boldsymbol{k})\epsilon_\mu^{(h)*}(\boldsymbol{k})\, e^{ikx} \right], \tag{8.34}$$

where

$$E_k = \sqrt{|\boldsymbol{k}|^2 + e^2 v^2}. \tag{8.35}$$

For each value of $\boldsymbol{k}$, the four-vectors $\epsilon^{(h)}(\boldsymbol{k})$ can be defined in the rest frame, where $k = (ev, 0)$, as the unit vectors in the four space-time directions:

$$\epsilon^{(0)}(\boldsymbol{k}) = \begin{pmatrix} 1 \\ 0 \\ 0 \\ 0 \end{pmatrix} \quad \epsilon^{(1)}(\boldsymbol{k}) = \begin{pmatrix} 0 \\ 1 \\ 0 \\ 0 \end{pmatrix} \quad \epsilon^{(2)}(\boldsymbol{k}) = \begin{pmatrix} 0 \\ 0 \\ 1 \\ 0 \end{pmatrix} \quad \epsilon^{(3)}(\boldsymbol{k}) = \begin{pmatrix} 0 \\ 0 \\ 0 \\ 1 \end{pmatrix}. \tag{8.36}$$

The corresponding expressions in a generic frame are obtained by suitable Lorentz transformations. In particular, we find $\epsilon_\mu^{(0)}(\boldsymbol{k}) = k_\mu/(ev)$; for this reason, the vector field component that corresponds to $h = 0$ is proportional to the four-divergence of a scalar field, and therefore describes a spin-0 degree of freedom.

It should be noted, however, that the Lagrangian density for A^0 has the opposite sign of those of the scalar fields H and G and of the space components A^i, $i = 1, 2, 3$. This is reflected in the sign of the propagator Eq. (8.32) for $\mu = \nu = 0$. This means that the spin-0 component of A does not correspond to physical quantum states, because it induces negative transition probabilities. Indeed, it can be shown that the creation and annihilation operators obey thecommutation rules

$$\left[a^{(h)}(\boldsymbol{k}), a^{\dagger(h')}(\boldsymbol{q}) \right] = -g^{hh'}\, \delta^{(3)}(\boldsymbol{k} - \boldsymbol{q}) \tag{8.37}$$

$$\left[a^{(h)}(\boldsymbol{k}), a^{(h')}(\boldsymbol{q}) \right] = \left[a^{\dagger(h)}(\boldsymbol{k}), a^{\dagger(h')}(\boldsymbol{q}) \right] = 0. \tag{8.38}$$

Note that the commutation rules for $a^{(h)}(\boldsymbol{k})$; $h = 1, 2, 3$ are the same as in the case of the real scalar field, while for $h = 0$ the sign of the commutator is reversed, as a consequence of the fact that the Lagrangian density for A^0 has the opposite sign of the Lagrangian for A^i, $i = 1, 2, 3$. This is easy to understand by analogy with the simple harmonic oscillator in ordinary quantum mechanics. In that case, the Hamiltonian is given by

$$H = \frac{p^2}{2m} + \frac{1}{2} m\omega^2 q^2, \tag{8.39}$$

and the creation and annihilation operators are related to p, q by

$$a = \sqrt{\frac{m\omega}{2}}\left(q + i\frac{1}{m\omega}p\right); \qquad a^{\dagger} = \sqrt{\frac{m\omega}{2}}\left(q - i\frac{1}{m\omega}p\right). \tag{8.40}$$

If the sign of the Lagrangian is reversed, $L \to -L$, then $p = dL/d\dot{q} \to -p$, and a and $a^{\dagger}$ are interchanged.

Let us now consider a one-particle state with polarization $h = 0$:

$$|\boldsymbol{k}\rangle = \int d^3p\, f_{\boldsymbol{k}}(\boldsymbol{p})\, a^{\dagger(0)}(\boldsymbol{p})|0\rangle, \tag{8.41}$$

where $|0\rangle$ is the vacuum state, and $f_{\boldsymbol{k}}(\boldsymbol{p})$ is a packet function, centered around $\boldsymbol{p} = \boldsymbol{k}$. Using Eq. (8.37), we find

$$\begin{aligned}\langle \boldsymbol{k}|\boldsymbol{k}\rangle &= \int d^3p\, d^3p'\, f_{\boldsymbol{k}}(\boldsymbol{p})\, f^{*}_{\boldsymbol{k}}(\boldsymbol{p}')\, \langle 0|a^{(0)}(\boldsymbol{p}')a^{\dagger(0)}(\boldsymbol{p})|0\rangle \\ &= -\langle 0|0\rangle \int d^3p\, |f_{\boldsymbol{k}}(\boldsymbol{p})|^2 .\end{aligned} \tag{8.42}$$

We see that the states generated by $a^{\dagger(0)}(\boldsymbol{p})$ have negative norm, and must therefore be removed from the physical spectrum. This can be achieved requiring

$$\epsilon^{(h)}(\boldsymbol{k}) \cdot k = 0; \quad \epsilon^{(h)}(\boldsymbol{k}) \cdot \epsilon^{(h)*}(\boldsymbol{k}) = -1 \tag{8.43}$$

for physical states. These conditions identify, for each value of the momentum, three independent polarization states, that correspond to the three different helicities of a spin-1 particle. In the rest frame of the vector boson, only the time component of k is non-zero, and ϵ must be a purely spatial vector, by Eq. (8.43). This excludes $\epsilon^{(0)}$ from the physical spectrum.

In practice, one is often interested in unpolarized cross sections, that are proportional to the tensor

$$P_{\mu\nu}(k) = \sum_{h=1}^{3} \epsilon^{(h)}_{\mu}(\boldsymbol{k})\, \epsilon^{(h)*}_{\nu}(\boldsymbol{k}). \tag{8.44}$$

It is easy to compute $P_{\mu\nu}$ in the rest frame of the vector boson: from Eq. (8.36) we get

$$P_{\mu\nu} = \begin{cases} 1 & \mu = \nu \neq 0 \\ 0 & \text{otherwise} \end{cases}, \tag{8.45}$$

or equivalently

$$P_{\mu\nu}(k) = -g_{\mu\nu} + \frac{k_{\mu}k_{\nu}}{e^2v^2}, \tag{8.46}$$

since $k = (ev, \boldsymbol{0})$ in the rest frame. Equation (8.46) is in covariant form, and therefore holds in any reference frame.

The problem of the internal quantum-mechanical consistency of the theory is not completely solved by imposing the constraints Eq. (8.43) on asymptotic states, because the scalar component of A^μ may still contribute to physical amplitudes as an intermediate state through the propagator, Eq. (8.32). However, it can be shown that such contributions are exactly cancelled by the intermediate states that correspond to the Goldstone field G, provided that one only considers processes in which the asymptotic states are either Higgs bosons, or spin-1 bosons. We will not prove this statement in full generality: we will illustrate how this cancellation takes place in a simple example. We consider the scattering process

$$H + \gamma \to H + \gamma. \tag{8.47}$$

The invariant amplitude can be computed by identifying the relevant Feynman diagrams, with two H and two A external lines. Recalling the Feynman rules given in the previous section, it is immediate to recognize that the relevant diagrams are

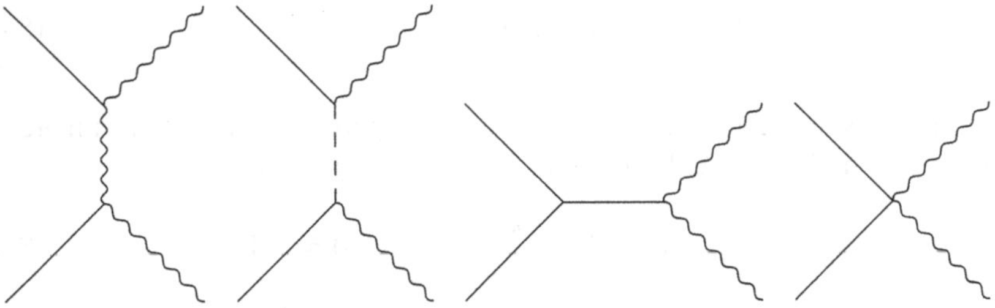

(A) (B) (C) (D)

Furthermore, one should include the diagrams obtained from (A) and (B) by permutation of the two external H lines:

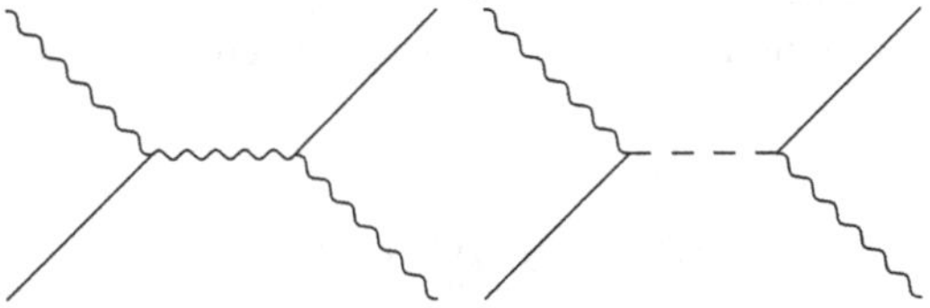

(A') (B')

Diagrams (A) and (A') contain a four-vector internal line, while diagrams (B) and (B') have a Goldstone internal line.

We now proceed to compute the amplitude. We assign momenta p and p' to initial and final H lines respectively, and momenta q and q' to initial and final state vector bosons. We denote by ϵ and ϵ' the corresponding polarization vectors, with the conditions

$$\epsilon \cdot q = \epsilon' \cdot q' = 0. \tag{8.48}$$

The invariant amplitude for diagrams (A) and (B) are given by

$$\mathcal{M}_A = \frac{4e^4 v^2 \epsilon \cdot \epsilon'}{(p+q)^2 - e^2 v^2} \tag{8.49}$$

$$\mathcal{M}_B = \frac{e^2 \epsilon \cdot (2p+q)\, \epsilon' \cdot (p+q+p')}{e^2 v^2 - (p+q)^2}. \tag{8.50}$$

Using Eq. (8.48) and four-momentum conservation $p+q = p'+q'$, we may replace $\epsilon \cdot (2p+q) \to 2\epsilon \cdot (p+q)$ and $\epsilon' \cdot p' \to \epsilon' \cdot (p+q)$. Thus,

$$\mathcal{M}_{A+B} = \frac{4e^4 v^2}{(p+q)^2 - e^2 v^2} \left[\epsilon \cdot \epsilon' - \frac{\epsilon \cdot (p+q)\, \epsilon' \cdot (p+q)}{e^2 v^2} \right]. \tag{8.51}$$

The amplitudes for diagrams (A') and (B') are obtained from $\mathcal{M}_{A+B}$ by the replacement $p \leftrightarrow -p'$:

$$\mathcal{M}_{A'+B'} = \frac{4e^4 v^2}{(p'-q)^2 - e^2 v^2} \left[\epsilon \cdot \epsilon' - \frac{\epsilon \cdot (p'-q)\, \epsilon' \cdot (p'-q)}{e^2 v^2} \right], \tag{8.52}$$

The expressions for $\mathcal{M}_{A+B}$ and $\mathcal{M}_{A'+B'}$, Eqs. (8.51, 8.52), are what we would obtain from diagrams (A) and (A') with the vector boson propagator replaced as follows:

$$\frac{g^{\mu\nu}}{q^2 - e^2 v^2 + i\epsilon} \to \frac{g^{\mu\nu} - \frac{q^\mu q^\nu}{e^2 v^2}}{q^2 - e^2 v^2 + i\epsilon}. \tag{8.53}$$

This corresponds to replacing $g^{\mu\nu}$ with the projection operator

$$g^{\mu\nu} - \frac{q^\mu q^\nu}{e^2 v^2}, \tag{8.54}$$

that projects a generic polarization vector on the physical subspace, defined by Eq. (8.43).

Finally, the amplitudes for diagrams (C) and (D) are given by

$$\mathcal{M}_C = -\frac{12\lambda e^2 v^2\, \epsilon \cdot \epsilon'}{2\lambda v^2 - (p-p')^2}; \qquad \mathcal{M}_D = 2\, e^2 \epsilon \cdot \epsilon'. \tag{8.55}$$

We observe that in this example the contributions of the scalar component of A and of the Goldstone degree of freedom G as intermediate states cancel exactly. It can be shown that this is a general feature of the theory: the abelian Higgs model is internally consistent, provided the asymptotic particles are either Higgs scalars, or spin-1 vectors. This result generalizes to more complicated Higgs models of relevance in the study of weak interactions.

This example shows that gauge invariance is needed in order to decouple non-physical from physical degrees of freedom, thus preserving unitarity. It is therefore necessary to prove that the counter-terms needed to make the theory finite when loop corrections are included do not violate gauge-invariance. It turns out that this is indeed the case.

In the semi-classical limit, non-physical degrees of freedom can be eliminated from the very beginning. We consider the first three terms of the Lagrangian density Eq. (8.22), that are invariant under gauge transformations, and we parametrize the scalar field as

$$\Phi(x) + \frac{v}{\sqrt{2}} = e^{ie\Theta(x)} \frac{H(x) + v}{\sqrt{2}}. \tag{8.56}$$

(instead of Eq. (8.24)). The field $\Theta(x)$ can be eliminated by a suitable gauge transformation:

$$\mathcal{A}_\mu(x) = A_\mu(x) - \partial_\mu \Theta(x). \tag{8.57}$$

In this way, we obtain an expression of the Lagrangian density that depends only on H and $\mathcal{A}$:

$$\begin{aligned} \mathcal{L}'_H = &\frac{1}{2}\partial_\mu H \partial^\mu H + \frac{e^2}{2}\mathcal{A}^2 (H+v)^2 - \frac{\lambda}{4}\left[(H+v)^2 - v^2\right]^2 \\ &- \frac{1}{2}\partial_\mu \mathcal{A}_\nu \left(\partial^\mu \mathcal{A}^\nu - \partial^\nu \mathcal{A}^\mu\right). \end{aligned} \tag{8.58}$$

The free Lagrangian density is now

$$\mathcal{L}_0 = \frac{1}{2}\left[(\partial H)^2 - 2\lambda v^2 H^2 - \partial_\mu \mathcal{A}_\nu \left(\partial^\mu \mathcal{A}^\nu - \partial^\nu \mathcal{A}^\mu\right) + e^2 v^2 \mathcal{A}^2\right], \tag{8.59}$$

and the field equations become

$$\left(\partial^2 + 2\lambda v^2\right) H = J_H \tag{8.60}$$

$$\left(\partial^2 + e^2 v^2\right) \mathcal{A}^\mu - \partial^\mu \partial_\nu \mathcal{A}^\nu = -J^\mu. \tag{8.61}$$

It is easy to show that the Green function for Eq. (8.61) corresponds to the propagator in the r.h.s. of Eq. (8.53). Thus, we find that, with the new gauge choice, only diagrams A, A', C and D contribute to the process in Eq. (8.47). However the amplitude does not change due to the change of the vector propagator.

8.4 The Higgs Mechanism in the Standard Model

The procedure described in Sect. 8.3 can be extended to the standard model, with few modifications. To this purpose, we introduce a scalar field that transforms non-trivially under that subset of gauge transformations that we want to undergo spontaneous breaking. We should keep in mind that the subgroup $U(1)_{\rm em}$ of the gauge group, that corresponds to electrodynamics, should not be spontaneously broken. In other words, we should perform the spontaneous breaking of the gauge symmetry in such a way that a mass term for the photon is not generated. This means that spontaneous symmetry breaking must take place in three of the four "directions" of the $SU(2) \times U(1)$ gauge group, the fourth one being that corresponding to electric charge. The simplest way to do this is to assign the scalar field ϕ to a doublet representation of $SU(2)$:

$$\phi = \begin{pmatrix} \phi_1 \\ \phi_2 \end{pmatrix} \tag{8.62}$$

with the $SU(2)$ transformation property

$$\phi \to e^{ig\alpha_i \tau_i/2} \left(\phi + \frac{v}{\sqrt{2}} \right) - \frac{v}{\sqrt{2}}, \tag{8.63}$$

where

$$v = \begin{pmatrix} v_1 \\ v_2 \end{pmatrix} \tag{8.64}$$

is a constant $SU(2)$ doublet. The value of the hypercharge of the scalar doublet ϕ is fixed by the requirement that the zero-field configuration is left unchanged by electromagnetic gauge transformations that correspond to the subgroup $U(1)_{\mathrm{em}}$, or equivalently that it be electrically neutral. This amounts to requiring that

$$e^{ie\alpha Q} \frac{1}{\sqrt{2}} \begin{pmatrix} v_1 \\ v_2 \end{pmatrix} = \frac{1}{\sqrt{2}} \begin{pmatrix} v_1 \\ v_2 \end{pmatrix}, \tag{8.65}$$

or equivalently

$$\begin{pmatrix} Q_1 & 0 \\ 0 & Q_2 \end{pmatrix} \begin{pmatrix} v_1 \\ v_2 \end{pmatrix} = \begin{pmatrix} 1/2 + Y/2 & 0 \\ 0 & -1/2 + Y/2 \end{pmatrix} \begin{pmatrix} v_1 \\ v_2 \end{pmatrix} = \begin{pmatrix} 0 \\ 0 \end{pmatrix}, \tag{8.66}$$

where Q_1, Q_2 are the electric charges of ϕ_1, ϕ_2, and we have used Eq. (7.36). The non-trivial solutions of Eq. (8.66) are

$$(1) \quad v_1 = 0, \quad |v_2| = v, \quad Y = +1 \tag{8.67}$$

$$(2) \quad v_2 = 0, \quad |v_1| = v, \quad Y = -1. \tag{8.68}$$

Without loss of generality, we will adopt the first choice, with $Y = +1$ and therefore $Q_1 = 1$, $Q_2 = 0$. We shall further assume that v_2 is real and positive.

The Higgs mechanism takes place in analogy with scalar electrodynamics. The most general scalar potential consistent with gauge invariance and renormalizability which is minimum at the origin of the isotopic space is

$$V(\phi) = \lambda\left[\left(\phi + \frac{v}{\sqrt{2}}\right)^{\dagger}\left(\phi + \frac{v}{\sqrt{2}}\right) - \frac{v^2}{2}\right]^2. \tag{8.69}$$

We can reparametrize ϕ in the following way:

$$\phi + \frac{v}{\sqrt{2}} = \frac{1}{\sqrt{2}} e^{i\tau^i\theta^i(x)/v}\begin{pmatrix} 0 \\ v + H(x)\end{pmatrix}, \tag{8.70}$$

with $\theta^i(x)$ and $H(x)$ real. With this parametrization it is apparent that the fields θ_i can be transformed away by an $SU(2)$ gauge transformation. As in the case of the abelian Higgs model. This choice, which is called the *unitary gauge*, is perfectly adequate for calculations in the semi-classical limit. However, it must be abandoned beyond this limit, as we shall see in Chap. 12. Here we will set $\theta_i = 0$.

The scalar potential takes the form

$$V = \frac{1}{2}(2\lambda v^2)H^2 + \lambda v H^3 + \frac{1}{4}\lambda H^4; \tag{8.71}$$

we see that the Higgs scalar H has a squared mass

$$m_H^2 = 2\lambda v^2. \tag{8.72}$$

Using Eq. (8.70) with $\theta^i = 0$ we get

$$\begin{aligned} D^\mu\left(\phi + \frac{v}{\sqrt{2}}\right) &= \left(\partial^\mu - i\frac{g}{2}\tau^i W^i_\mu - i\frac{g'}{2}B_\mu\right)\frac{1}{\sqrt{2}}\begin{pmatrix} 0 \\ H(x) + v\end{pmatrix} \\ &= \frac{1}{\sqrt{2}}\begin{pmatrix} 0 \\ \partial^\mu H\end{pmatrix} - \frac{i}{2}(H+v)\frac{1}{\sqrt{2}}\begin{pmatrix} g(W_1^\mu - iW_2^\mu) \\ -gW_3^\mu + g'B^\mu\end{pmatrix} \\ &= \frac{1}{\sqrt{2}}\begin{pmatrix} 0 \\ \partial^\mu H\end{pmatrix} - \frac{i}{2}(H+v)\begin{pmatrix} gW^{\mu+} \\ -\sqrt{(g^2+g'^2)/2}Z^\mu\end{pmatrix}, \end{aligned} \tag{8.73}$$

where in the last step we have used Eqs. (7.20), (7.27), (7.28) and (7.35). We have therefore

$$\begin{aligned} &\left[D^\mu\left(\phi + \frac{v}{\sqrt{2}}\right)\right]^{\dagger} D_\mu\left(\phi + \frac{v}{\sqrt{2}}\right) \\ &= \frac{1}{2}\partial^\mu H\partial_\mu H + \left[\frac{1}{4}g^2 W^{\mu+}W^-_\mu + \frac{1}{8}(g^2+g'^2)Z^\mu Z_\mu\right](H+v)^2. \end{aligned} \tag{8.74}$$

We see that the W and Z bosons have acquired masses

$$m_W^2 = \frac{1}{4} g^2 v^2 \tag{8.75}$$

$$m_Z^2 = \frac{1}{4} (g^2 + g'^2) v^2. \tag{8.76}$$

Note that the photon remains massless. With the scalar field ϕ transforming as a doublet of $SU(2)$, there is always a linear combination of B^μ and W_3^μ that does not receive a mass term, but only if $Y(\phi) = 1$ (or -1) does this linear combination coincide with the one in Eq. (7.27).

The value of v, the *vacuum expectation value* of the neutral component of the Higgs doublet, can be obtained combining Eqs. (8.3) and (8.75), and using the measured valued of the Fermi constant. We get

$$v = \sqrt{\frac{1}{G_F \sqrt{2}}} \simeq 246.22 \text{ GeV}. \tag{8.77}$$

The value of the Higgs quartic coupling λ can be obtained from Eq. (8.72), using the measured value of the Higgs boson mass, $m_H = 125$ GeV. We get

$$\lambda = \frac{m_H^2}{2v^2} \simeq 0.13. \tag{8.78}$$

Chapter 9
Breaking of Accidental Symmetries

9.1 Quark Masses and Flavour-Mixing

Fermion mass terms are forbidden by the gauge symmetry of the standard model. Indeed, a Dirac mass term for a fermion field ψ

$$-m\,\overline{\psi}\psi = -m\left(\overline{\psi}_L\psi_R + \overline{\psi}_R\psi_L\right) \tag{9.1}$$

is not invariant under a chiral transformation, i.e. a transformation that acts differently on left-handed and right-handed components. The gauge transformations of the standard model are precisely of this kind. Again, this difficulty can be circumvented by means of the Higgs mechanism.

We first consider the hadronic sector. We have seen in Sect. 7.3 that the interaction Lagrangian is not diagonal in terms of quark fields with definite flavours. Let u'^f and d'^f be the fields that bring the interaction terms in diagonal form (the index f runs over the n fermion generations); in principle, there is no reason why only down-type quark fields should be rotated. We also define

$$q'^f_L = \begin{pmatrix} u'^f_L \\ d'^f_L \end{pmatrix}. \tag{9.2}$$

A Yukawa interaction term can be added to the Lagrangian density:

$$\begin{aligned}\mathcal{L}_Y^{\rm hadr} = &-\left[\bar{q}'_L\left(\phi+\frac{v}{\sqrt{2}}\right)h'_{\rm D}\,d'_R + \bar{d}'_R\left(\phi+\frac{v}{\sqrt{2}}\right)^{\dagger}h'^{\dagger}_{\rm D}\,q'_L\right]\\ &-\left[\bar{q}'_L\left(\phi+\frac{v}{\sqrt{2}}\right)_c h'_{\rm U}\,u'_R + \bar{u}'_R\left(\phi+\frac{v}{\sqrt{2}}\right)^{\dagger}_c h'^{\dagger}_{\rm U}\,q'_L\right],\end{aligned} \tag{9.3}$$

C. M. Becchi and G. Ridolfi, *An Introduction to Relativistic Processes and the Standard Model of Electroweak Interactions*, UNITEXT for Physics,
DOI: 10.1007/978-3-319-06130-6_9, © Springer International Publishing Switzerland 2014

where h'_U and h'_D are generic $n \times n$ constant complex matrices in the generation space. The scalar doublet

$$\left(\phi + \frac{v}{\sqrt{2}}\right)_c = \epsilon \left(\phi + \frac{v}{\sqrt{2}}\right)^*, \tag{9.4}$$

where ϵ is the antisymmetric matrix in two dimensions, can be shown to transform as a doublet under $SU(2)_L$.

It easy to check that $\mathcal{L}_Y^{\rm hadr}$ is Lorentz-invariant, gauge-invariant and renormalizable, and therefore it can be included in the Lagrangian. The matrices h'_U and h'_D can be diagonalized by means of bi-unitary transformations:

$$h_U \equiv V_L^{U\dagger} h'_U V_R^U \tag{9.5}$$

$$h_D \equiv V_L^{D\dagger} h'_D V_R^D, \tag{9.6}$$

where $V_{L,R}^{U,D}$ are unitary matrices, chosen so that h_U and h_D are diagonal with real, non-negative entries. Now, we define new quark fields u and d by

$$u'_L = V_L^U u_L, \quad u'_R = V_R^U u_R \tag{9.7}$$

$$d'_L = V_L^D d_L, \quad d'_R = V_R^D d_R, \tag{9.8}$$

In the unitary gauge, Eq. (9.3) becomes

$$\mathcal{L}_Y^{\rm hadr} = -\frac{1}{\sqrt{2}}(v + H) \sum_{f=1}^{n} (h_D^f \bar{d}^f d^f + h_U^f \bar{u}^f u^f), \tag{9.9}$$

where $h_{U,D}^f$ are the diagonal entries of the matrices $h_{U,D}$. We can now identify the quark masses with

$$m_U^f = \frac{v h_U^f}{\sqrt{2}}, \quad m_D^f = \frac{v h_D^f}{\sqrt{2}}. \tag{9.10}$$

Since the matrices $V_{L,R}^{U,D}$ are constant in space-time, Eqs. (9.7) and (9.8) obviously define global symmetry transformations of the free quark Lagrangian. They also leave unchanged the neutral-current interaction term, because of the universality of the couplings of fermions of different families to the photon and the Z. The only term in the Lagrangian which is affected by the transformations in Eqs. (9.7) and (9.8) is the charged-current interaction, because the up and down components of the same left-handed doublet are transformed in different ways. Indeed, we find

$$\mathcal{L}_C^{\rm hadr} = \frac{g}{\sqrt{2}} W_\mu^+ \sum_{f=1}^{n} \bar{q}'^f_L \gamma^\mu \tau^+ q'^f_L + \text{h.c.} = \frac{g}{\sqrt{2}} W_\mu^+ \sum_{f,g=1}^{n} \bar{u}_L^f \gamma^\mu V_{fg} d_L^g + \text{h.c.}, \tag{9.11}$$

where

$$V = V_L^{U\dagger} V_L^D. \tag{9.12}$$

The matrix V is usually called the *Cabibbo-Kobayashi-Maskawa* (CKM) matrix. It is a unitary matrix, and its unitarity guarantees the suppression of flavour-changing neutral currents, as we already discussed in Sect. 7.1 in the case of two fermion families. The elements of V are fundamental parameters of the standard model Lagrangian, on the same footing as masses and gauge couplings, and must be extracted from experiments.

To conclude this Section, we determine the number of independent parameters in the CKM matrix. A generic $n \times n$ unitary matrix depends on n^2 independent real parameters. Some (n_A) of them can be thought of as rotation angles in the n-dimensional space of generations, and they are as many as the coordinate planes in n dimensions:

$$n_A = \binom{n}{2} = \frac{1}{2}n(n-1). \tag{9.13}$$

The remaining

$$\hat{n}_P = n^2 - n_A = \frac{1}{2}n(n+1) \tag{9.14}$$

parameters are complex phases. Some of them can be eliminated from the Lagrangian density by redefining the left-handed quark fields as

$$u_L^f \to e^{i\alpha_f}\, u_L^f; \quad d_L^g \to e^{i\beta_g}\, d_L^g, \tag{9.15}$$

with α_f, β_g real constants. Indeed, the transformations Eq. (9.15) are symmetry transformations for the whole standard model Lagrangian except $\mathcal{L}_c^{\text{hadr}}$, and therefore amount to a redefinition of the CKM matrix:

$$V_{fg} \to e^{i(\beta_g - \alpha_f)}\, V_{fg}. \tag{9.16}$$

The $2n$ constants α_f, β_g can be chosen so that $2n-1$ phases are eliminated from the matrix V, since there are $2n-1$ independent differences $\beta_g - \alpha_f$. The number of really independent complex phases in V is therefore

$$n_P = \hat{n}_P - (2n-1) = \frac{1}{2}(n-1)(n-2). \tag{9.17}$$

Observe that, with one or two fermion families, the CKM matrix can be made real. The first case with non-trivial phases is $n = 3$, which corresponds to $n_P = 1$. In the standard model with three fermion families, the CKM matrix has four independent parameters: three rotation angles and one complex phase. In the general case, the total number of independent parameters in the CKM matrix is

$$n_A + n_P = (n-1)^2. \tag{9.18}$$

The presence of complex coupling constants implies violation of the CP symmetry. CP violation phenomena in weak interactions were first observed around 1964 in the $K^0 - \overline{K}^0$ system; the existence of a third quark generation may therefore be considered as a prediction of the standard model, confirmed by the discovery of the b and t quarks.

9.2 Lepton Masses

The same procedure can be applied to the leptonic sector. Everything is formally unchanged: up-quarks are replaced by neutrinos and down-quarks are replaced by charged leptons (e^-, μ^- and τ^-). There is however an important difference, which leads to considerable simplifications: as we have seen, right-handed neutrinos have no interactions, and hence can be omitted. Therefore, there is no need to introduce a Yukawa coupling involving the conjugate scalar field ϕ_c:

$$\mathcal{L}_Y^{\text{lept}} = -\left[\bar{\ell}'_L \left(\phi + \frac{v}{\sqrt{2}}\right) h'_{\text{E}}\, e'_R + \bar{e}'_R \left(\phi + \frac{v}{\sqrt{2}}\right)^\dagger h'^\dagger_{\text{E}}\, \ell'_L\right]. \tag{9.19}$$

The matrix of Yukawa couplings h'_{E} can be diagonalized by means of a bi-unitary transformation

$$h_{\text{E}} = V_{\text{L}}^{\text{E}\dagger} h'_{\text{E}} V_{\text{R}}^{\text{E}}. \tag{9.20}$$

The difference with respect to the case of quarks is that now we have the freedom of redefining the left-handed neutrino fields using *the same* matrix V_{L}^{E} that rotates charged leptons:

$$\nu'_L = V_{\text{L}}^{\text{E}} \nu_L \tag{9.21}$$

$$e'_L = V_{\text{L}}^{\text{E}} e_L, \quad e'_R = V_{\text{R}}^{\text{E}} e_R. \tag{9.22}$$

This brings the lepton Yukawa interaction in diagonal form,

$$\mathcal{L}_Y^{\text{lept}} = -\sum_{f=1}^{n} h_{\text{E}}^f \left[\bar{\ell}_L^f \left(\phi + \frac{v}{\sqrt{2}}\right) e_R^f + \bar{e}_R^f \left(\phi + \frac{v}{\sqrt{2}}\right)^\dagger \ell_L^f\right], \tag{9.23}$$

but, contrary to what happens in the quark sector, leaves the charged-current interaction term unchanged:

$$\mathcal{L}_c^{\text{lept}} = \frac{g}{\sqrt{2}} W_\mu^+ \sum_{f=1}^{n} \bar{\ell}'^f_L \gamma^\mu \tau^+ \ell'^f_L + \text{h.c.} = \frac{g}{\sqrt{2}} W_\mu^+ \sum_{f=1}^{n} \bar{\ell}^f_L \gamma^\mu \tau^+ \ell^f_L + \text{h.c.}. \tag{9.24}$$

In other words, in the leptonic sector there is no mixing among different generations, because the Yukawa coupling matrix can be diagonalized by a global transformation under which the full Lagrangian is invariant. As a consequence, not only the overall leptonic number, but also individual leptonic flavours are conserved. This is due to the absence of right-handed neutrinos (see however Chap. 13).

The values of the Yukawa couplings h_{E}^{f} are determined by the values of the observed lepton masses. In fact, using Eq. (8.70), we find

$$\mathcal{L}_{Y}^{\mathrm{lept}} = -\sum_{f=1}^{n} \frac{h_{\mathrm{E}}^{f}}{\sqrt{2}}(v + H)\, \bar{e}_f\, e_f, \tag{9.25}$$

thus allowing the identifications

$$m_{\mathrm{E}}^{f} = \frac{v h_{\mathrm{E}}^{f}}{\sqrt{2}}. \tag{9.26}$$

9.3 Accidental Symmetries

The need for Yukawa interaction terms of fermion fields with scalar fields can be motivated in a different way. Consider the standard model with only one generation of quarks and leptons, and no scalar fields. The Lagrangian for fermion fields can be written in the following compact form:

$$\mathcal{L} = \sum_{k=1}^{5} \bar{\psi}_k\, i \not{D} \psi_k, \tag{9.27}$$

where the sum runs over the five different irreducible representations of $SU(2) \times U(1)$ of fermions within a generation:

$$\begin{aligned}
\psi_1 &= e_R \sim (\mathbf{1}, -2)\\
\psi_2 &= \ell_L \sim (\mathbf{2}, -1)\\
\psi_3 &= u_R \sim (\mathbf{1}, 4/3)\\
\psi_4 &= d_R \sim (\mathbf{1}, -2/3)\\
\psi_5 &= q_L \sim (\mathbf{2}, 1/3).
\end{aligned}$$

Here, the symbol $\sim$ means "transforms as", and the two numbers in brackets stand for the $SU(2)$ representation (**2** for the doublet, **1** for the scalar) and for the hypercharge quantum number, respectively. Mass terms are forbidden by the gauge symmetry.

In addition to the assumed gauge symmetry, the Lagrangian in Eq. (9.27) is manifestly invariant under a large class of global transformations: namely, the fermion fields within each representation can be multiplied by an arbitrary constant phase

$$\psi_k \to e^{i\phi_k}\psi_k \tag{9.28}$$

without affecting $\mathcal{L}$. This $[U(1)]^5$ global symmetry was not imposed at the beginning: it is just a consequence of the assumed gauge symmetry and of the renormalizability condition. It is therefore called an *accidental* symmetry.

Let us take a closer look at the accidental symmetry. The five conserved currents corresponding to the global transformations (9.28) are

$$\begin{aligned}
J_1^\mu &= \bar{e}_R\gamma^\mu e_R \\
J_2^\mu &= \bar{\nu}_L\gamma^\mu \nu_L + \bar{e}_L\gamma^\mu e_L \\
J_3^\mu &= \bar{u}_R\gamma^\mu u_R \\
J_4^\mu &= \bar{d}_R\gamma^\mu d_R \\
J_5^\mu &= \bar{u}_L\gamma^\mu u_L + \bar{d}_L\gamma^\mu d_L .
\end{aligned}$$

A deeper insight in the meaning of these conserved currents is achieved by replacing $J_1^\mu, \ldots, J_5^\mu$ by the following independent linear combinations of them:

$$\begin{aligned}
J_Y^\mu &= \sum_{k=1}^{5} \frac{Y_k}{2} J_k^\mu \\
J_\ell^\mu &= J_1^\mu + J_2^\mu = \bar{\nu}\gamma^\mu\nu + \bar{e}\gamma^\mu e \\
J_{\ell 5}^\mu &= J_1^\mu - J_2^\mu = \bar{\nu}\gamma^\mu\gamma_5\nu + \bar{e}\gamma^\mu\gamma_5 e \\
J_b^\mu &= \frac{1}{3}(J_3^\mu + J_4^\mu + J_5^\mu) = \frac{1}{3}(\bar{u}\gamma^\mu u + \bar{d}\gamma^\mu d) \\
J_{b5}^\mu &= J_3^\mu + J_4^\mu - J_5^\mu = \bar{u}\gamma^\mu\gamma_5 u + \bar{d}\gamma^\mu\gamma_5 d .
\end{aligned}$$

The current J_Y is the hypercharge current, which corresponds to a local invariance of the theory. The true accidental symmetry is therefore $[U(1)]^4$, rather than $[U(1)]^5$.

The currents J_ℓ and J_b are immediately recognized to be the leptonic and baryonic number currents, respectively. The invariance of the Lagrangian under the corresponding global symmetries is certainly good news, since baryonic and leptonic number are known to be conserved to an extremely high accuracy. On the other hand, experiments show no sign of the conservation of $J_{\ell 5}$ and J_{b5}; in a realistic theory, the corresponding symmetries should be broken. In fact, they are incompatible with mass terms, and they are broken by the Yukawa interaction terms that generate fermion masses via the Higgs mechanism. When the theory is extended to include more fermion generations, the accidental symmetry gets much larger, since also mixing among different generations is allowed. The Yukawa interaction terms of the previous Sections break this larger accidental symmetry too, leaving however baryonic and leptonic numbers conserved. Individual leptonic numbers are separately conserved, while only the total baryonic number is conserved, because of flavour mixing in the quark sector.

It should be noted that, because of accidental symmetries, Yukawa interactions cannot be generated by radiative corrections. This is the mechanism that keeps neutrinos massless, and that protects fermion masses from receiving large radiative corrections.

We conclude this Section by reviewing the most important experimental evidences of baryon and lepton number conservation. The most obvious test of baryon number conservation is proton stability. The experimental lower bound on the proton lifetime is at present

$$\tau_p > 2.1 \times 10^{29}\ y. \tag{9.29}$$

The most accurate tests of lepton number conservation are provided by the following observables:

$$B(\mu \to e\gamma) \leq 1.2 \times 10^{-11}; \qquad B(\tau \to \mu\gamma) \leq 2.7 \times 10^{-6} \tag{9.30}$$

$$B(\mu \to 3e) \leq 1 \times 10^{-12} \tag{9.31}$$

$$\frac{\Gamma(\mu\ \mathrm{Ti} \to e\ \mathrm{Ti})}{\Gamma(\mu\ \mathrm{Ti} \to \mathrm{all})} \leq 4 \times 10^{-12}, \tag{9.32}$$

where B stands for the ratio between the rate of the indicated process and the total decay rate.

Chapter 10
Summary

10.1 The Standard Model Lagrangian in the Unitary Gauge

We present here the full Lagrangian density of the standard model with one Higgs doublet. We have

$$\mathcal{L}_{\rm SM} = \mathcal{L}_0^F + \mathcal{L}_0^G + \mathcal{L}_{\rm em} + \mathcal{L}_c + \mathcal{L}_n + \mathcal{L}_{\rm V} + \mathcal{L}_{\rm Higgs} \tag{10.1}$$

where

- $\mathcal{L}_0^F$ is the free Lagrangian for matter fermions:

$$\mathcal{L}_0^F = \sum_{f=1}^{n} \left[\bar{\nu}^f i\partial\!\!\!/ \nu^f + \bar{e}^f (i\partial\!\!\!/ - m_{\rm E}^f) e^f + \bar{u}^f (i\partial\!\!\!/ - m_{\rm U}^f) u^f + \bar{d}^f (i\partial\!\!\!/ - m_{\rm D}^f) d^f \right]. \tag{10.2}$$

  The index f labels the n fermion families.

- $\mathcal{L}_0^G$ is the free Lagrangian for gauge and Higgs bosons:

$$\begin{aligned}\mathcal{L}_0^G = &-\frac{1}{4} Z_{\mu\nu} Z^{\mu\nu} + \frac{1}{2} m_Z^2 Z^\mu Z_\mu - \frac{1}{2} W^+_{\mu\nu} W^{\mu\nu}_- + m_W^2 W^{\mu+} W^-_\mu \\ &- \frac{1}{2} \partial_\mu A_\nu \partial^\mu A^\nu + \frac{1}{2} \partial^\mu H \partial_\mu H - \frac{1}{2} m_H^2 H^2,\end{aligned} \tag{10.3}$$

  where

$$Z^{\mu\nu} = \partial^\mu Z^\nu - \partial^\nu Z^\mu; \quad W_\pm^{\mu\nu} = \partial^\mu W_\pm^\nu - \partial^\nu W_\pm^\mu. \tag{10.4}$$

C. M. Becchi and G. Ridolfi, *An Introduction to Relativistic Processes and the Standard Model of Electroweak Interactions*, UNITEXT for Physics,
DOI: 10.1007/978-3-319-06130-6_10, © Springer International Publishing Switzerland 2014

- $\mathcal{L}_{\text{em}}$ is the electromagnetic coupling:

$$\mathcal{L}_{\text{em}} = e \sum_{f=1}^{n} \left(-\bar{e}^f \gamma_\mu e^f + \frac{2}{3} \bar{u}^f \gamma_\mu u^f - \frac{1}{3} \bar{d}^f \gamma_\mu d^f \right) A^\mu. \tag{10.5}$$

- $\mathcal{L}_c$ is the charged-current interaction term:

$$\begin{aligned}\mathcal{L}_c = &\frac{g}{2\sqrt{2}} \left[\sum_{f=1}^{n} \bar{\nu}^f \gamma^\mu (1-\gamma_5) e^f + \sum_{f,g=1}^{n} \bar{u}^f \gamma^\mu (1-\gamma_5) V_{fg} d^g \right] W_\mu^+ \\ &+ \frac{g}{2\sqrt{2}} \left[\sum_{f=1}^{n} \bar{e}^f \gamma^\mu (1-\gamma_5) \nu^f + \sum_{f,g=1}^{n} \bar{d}^g \gamma^\mu (1-\gamma_5) V_{fg}^* u^f \right] W_\mu^-. \end{aligned} \tag{10.6}$$

- $\mathcal{L}_n$ is the neutral-current interaction term:

$$\begin{aligned}\mathcal{L}_n = &\frac{e}{4\cos\theta_W \sin\theta_W} \sum_{f=1}^{n} \left[\bar{\nu}^f \gamma_\mu (1-\gamma_5) \nu^f + \bar{e}^f \gamma_\mu \left(-1 + 4\sin^2\theta_W + \gamma_5 \right) e^f \right. \\ &\left. + \bar{u}^f \gamma_\mu \left(1 - \frac{8}{3}\sin^2\theta_W - \gamma_5 \right) u^f + \bar{d}^f \gamma_\mu \left(-1 + \frac{4}{3}\sin^2\theta_W + \gamma_5 \right) d^f \right] Z^\mu. \end{aligned} \tag{10.7}$$

- $\mathcal{L}_V$ contains vector boson interactions among themselves:

$$\begin{aligned}\mathcal{L}_V = &+ig\sin\theta_W (W_{\mu\nu}^+ W_-^\mu A^\nu - W_{\mu\nu}^- W_+^\mu A^\nu + F_{\mu\nu} W_+^\mu W_-^\nu) \\ &+ ig\cos\theta_W (W_{\mu\nu}^+ W_-^\mu Z^\nu - W_{\mu\nu}^- W_+^\mu Z^\nu + Z_{\mu\nu} W_+^\mu W_-^\nu) \\ &+ \frac{g^2}{2} (2g^{\mu\nu} g^{\rho\sigma} - g^{\mu\rho} g^{\nu\sigma} - g^{\mu\sigma} g^{\nu\rho}) \left[\frac{1}{2} W_\mu^+ W_\nu^+ W_\rho^- W_\sigma^- \right. \\ &\left. - W_\mu^+ W_\nu^- (A_\rho A_\sigma \sin^2\theta_W + Z_\rho Z_\sigma \cos^2\theta_W + 2A_\rho Z_\sigma \sin\theta_W \cos\theta_W) \right] \end{aligned} \tag{10.8}$$

where

$$F^{\mu\nu} = \partial^\mu A^\nu - \partial^\nu A^\mu. \tag{10.9}$$

- The Higgs interaction Lagrangian is given by

$$\mathcal{L}_{\text{Higgs}} = \left(m_W^2 W^{\mu+} W_\mu^- + \frac{1}{2} m_Z^2 Z^\mu Z_\mu\right)\left(\frac{H^2}{v^2} + \frac{2H}{v}\right)$$
$$- \frac{H}{v} \sum_{f=1}^{n} (m_D^f \bar{d}^f d^f + m_U^f \bar{u}^f u^f + m_E^f \bar{e}^f e^f)$$
$$- \lambda v H^3 - \frac{1}{4}\lambda H^4. \tag{10.10}$$

We have adopted the gauge fixing Eq. (6.38) for electrodynamics, so that the photon propagator is

$$\Delta_\gamma^{\mu\nu}(k) = \frac{g^{\mu\nu}}{k^2 + i\epsilon}, \tag{10.11}$$

and the unitary gauge for weak interactions, so that the W and Z boson propagators are

$$\Delta_V^{\mu\nu}(k) = \frac{1}{k^2 - m_V^2}\left(g^{\mu\nu} - \frac{k^\mu k^\nu}{m_V^2}\right); \qquad V = W, Z. \tag{10.12}$$

10.2 The Standard Model Lagrangian in Renormalizable Gauges

The unitary gauge is not suited for calculations beyond the semi-classical approximation. In such cases, manifest renormalizability is of great importance, even if the price to pay is the inclusion of non-physical degrees of freedom in internal lines of Feynman diagrams.

In this section, we illustrate how the standard model Lagrangian of the previous section is modified if the 't Hooft-Feynman gauge-fixing term is chosen:

$$\mathcal{L}_{\text{gf}} = -\frac{1}{2}\left[\partial^\mu W_\mu^i - f^i(\phi_1, \phi_2)\right]^2 - \frac{1}{2}\left[\partial^\mu B_\mu - f(\phi_1, \phi_2)\right]^2$$
$$+ \partial^\mu \bar{c}^i \left(\partial_\mu c^i - g\epsilon^{ijk} c^j W_\mu^k\right) + \partial^\mu \bar{c}^0 \partial_\mu c^0$$
$$- \frac{g\bar{c}^i}{4}\left[\phi_2^\dagger (gc^j \tau^j + g' c^0)\tau^i \phi_1 + \phi_1^\dagger (gc^j \tau^j + g' c^0)\tau^i \phi_2\right]$$
$$- \frac{g'\bar{c}^0}{4}\left[\phi_2^\dagger (gc^j \tau^j + g' c^0)\phi_1 + \phi_1^\dagger (gc^j \tau^j + g' c^0)\phi_2\right] \tag{10.13}$$

where

$$\phi_1 = \frac{1}{\sqrt{2}}\begin{pmatrix} 0 \\ v \end{pmatrix} \qquad \phi_2 = \begin{pmatrix} G^+ \\ \frac{H + iG}{\sqrt{2}} \end{pmatrix} \tag{10.14}$$

$$f^i(\phi_1, \phi_2) = -\frac{i}{2}g(\phi_2^\dagger \tau^i \phi_1 - \phi_1^\dagger \tau^i \phi_2) \tag{10.15}$$

$$f(\phi_1, \phi_2) = -\frac{i}{2}g'(\phi_2^\dagger \phi_1 - \phi_1^\dagger \phi_2) \tag{10.16}$$

and c^i, c^0, $\bar{c}^i$, $\bar{c}^0$ are ghost and antighost fields, which are introduced following the Faddeeev-Popov quantization prescription. The Lagrangian of the previous section gets modified as follows:

- The free Lagrangian in the gauge-Higgs sector $\mathcal{L}_0^G$ becomes

$$\begin{aligned}\mathcal{L}_0^G = & -\frac{1}{2}\partial_\mu Z_\nu \partial^\mu Z^\mu + \frac{1}{2}m_Z^2 Z^\mu Z_\mu - \partial_\mu W_\nu^+ \partial^\mu W_-^\nu + m_W^2 W^{\mu+} W_\mu^- - \frac{1}{2}\partial_\mu A_\nu \partial^\mu A^\nu \\ & + \frac{1}{2}\partial^\mu H \partial_\mu H - \frac{1}{2}m_H^2 H^2 + \frac{1}{2}\partial^\mu G \partial_\mu G - \frac{1}{2}m_Z^2 G^2 \\ & + \partial^\mu G^+ \partial_\mu G^- - m_W^2 G^+ G^-. \end{aligned} \tag{10.17}$$

  Note that quadratic terms for the Goldston bosons G, $G^\pm$ have appeared, and that the derivative terms for the W_μ and Z_μ fields have the same form as that of A_μ.
- The Higgs interaction terms $\mathcal{L}_{\text{Higgs}}$ are also heavily modified with respect to the unitary gauge. We find

$$\begin{aligned}\mathcal{L}_{\text{Higgs}} = & \left(m_W^2 W^{\mu+} W_\mu^- + \frac{1}{2}m_Z^2 Z^\mu Z_\mu\right)\left(\frac{H^2}{v^2} + \frac{2H}{v}\right) \\ & + \frac{1}{2}g^2 W^{+\mu} W_\mu^- (G^+ G^- + \frac{1}{2}G^2) + \frac{1}{8}\frac{g^2}{\cos^2\theta_W} Z^\mu Z_\mu G^2 \\ & + e m_W (A^\mu - \tan\theta_W Z^\mu)(G^+ W_\mu^- + G^- W_\mu^+) \\ & - \frac{ig}{2} W_\mu^+ (H + iG)(\partial^\mu + ieA^\mu - ie\tan\theta_W Z^\mu) G^- \\ & + \frac{ig}{2} W_\mu^- (H - iG)(\partial^\mu - ieA^\mu + ie\tan\theta_W Z^\mu) G^+ \\ & + \frac{ig}{2}[G^- W_\mu^+ \partial^\mu (H + iG) - G^+ W_\mu^- \partial^\mu (H - iG)] \\ & - \frac{g}{2\cos\theta_W}(G\partial_\mu H - H\partial_\mu G) Z^\mu \\ & - i\left[eA^\mu + \frac{g\cos 2\theta_W}{2\cos\theta_W} Z^\mu\right](G^+ \partial_\mu G^- - G^- \partial_\mu G^+) \\ & + \left[eA^\mu + \frac{g\cos 2\theta_W}{2\cos\theta_W} Z^\mu\right]^2 G_+ G_- \\ & - \lambda\left[vH(H^2 + 2G^+ G^- + G^2) + \frac{\left(H^2 + 2G^+ G^- + G^2\right)^2}{4}\right] \\ & - \frac{H}{v}(\bar{d}m_{\text{D}}d + \bar{u}m_{\text{U}}u + \bar{e}m_{\text{E}}e) - \frac{iG}{v}(\bar{d}m_{\text{D}}\gamma_5 d - \bar{u}m_{\text{U}}\gamma_5 u + \bar{e}m_{\text{D}}\gamma_5 e)\end{aligned}$$

$$
\begin{aligned}
&-\frac{\sqrt{2}}{v}G^{+}\left(\bar{u}_L V m_{\mathrm{D}} d_R - \bar{u}_R m_{\mathrm{U}} V d_L\right) - \frac{\sqrt{2}}{v}G^{-}\left(\bar{d}_R m_{\mathrm{D}} V^{\dagger} u_L - \bar{d}_L V^{\dagger} m_{\mathrm{U}} u_R\right) \\
&-\frac{\sqrt{2}}{v}G^{+}\bar{\nu}_L m_{\mathrm{E}} e_R - \frac{\sqrt{2}}{v}G^{-}\bar{e}_R m_{\mathrm{E}} \nu_L, \qquad (10.18)
\end{aligned}
$$

where sums over generation indices are understood.

- Setting

$$
\begin{aligned}
c^z &= \sin\theta_W c^0 - \cos\theta_W c^3, \quad & c^a &= \sin\theta_W c^3 + \cos\theta_W c^0, \quad & c^{\pm} &= \frac{c^1 \mp i c^2}{\sqrt{2}} \\
\bar{c}^z &= \sin\theta_W \bar{c}^0 - \cos\theta_W \bar{c}^3, \quad & \bar{c}^a &= \sin\theta_W \bar{c}^3 + \cos\theta_W \bar{c}^0, \quad & \bar{c}^{\pm} &= \frac{\bar{c}^1 \mp i \bar{c}^2}{\sqrt{2}}
\end{aligned} \qquad (10.19)
$$

one has the free Lagrangian of the ghosts

$$
\begin{aligned}
\mathcal{L}_{\mathrm{ghost},0} = {} & \partial^{\mu}\bar{c}^{+}\partial_{\mu}c^{-} + \partial^{\mu}\bar{c}^{-}\partial_{\mu}c^{+} + \partial^{\mu}\bar{c}^{z}\partial_{\mu}c^{z} + \partial^{\mu}\bar{c}^{a}\partial_{\mu}c^{a} \\
& - m_W^2(\bar{c}^{+}c^{-} + \bar{c}^{-}c^{+}) - m_Z^2\bar{c}^{z}c^{z} \qquad (10.20)
\end{aligned}
$$

and the ghost interaction Lagrangian

$$
\begin{aligned}
\mathcal{L}_{\mathrm{ghost},I} = {} & ieA^{\mu}(\partial_{\mu}\bar{c}^{+}c^{-} - \partial_{\mu}\bar{c}^{-}c^{+}) \\
& + ig\Big[\partial^{\mu}(\sin\theta_W\bar{c}^{a} + \cos\theta_W\bar{c}^{z})\left(c^{+}W_{\mu}^{-} - c^{-}W_{\mu}^{+}\right) \\
& + \cos\theta_W\left(\partial^{\mu}\bar{c}^{+}c^{-} - \partial^{\mu}\bar{c}^{-}c^{+}\right)Z_{\mu} \\
& + \left(\partial^{\mu}\bar{c}^{-}W_{\mu}^{+} - \partial^{\mu}\bar{c}^{+}W_{\mu}^{-}\right)(\sin\theta_W c^{a} + \cos\theta_W c^{z})\Big] \\
& - i\frac{m_W^2}{v}\left(\bar{c}^{-}c^{+} - \bar{c}^{+}c^{-}\right)G \\
& - \frac{H}{v}\left(m_W^2(\bar{c}^{+}c^{-} + \bar{c}^{-}c^{+}) + m_Z^2\bar{c}^{z}c^{z}\right) + \frac{m_W m_Z}{v}\bar{c}^{z}(c^{+}G^{-} + c^{-}G^{+}) \\
& - \frac{m_W m_Z}{v}(\bar{c}^{+}G^{-} + \bar{c}^{-}G^{+})(\sin(2\theta_W)c^{a} + \cos(2\theta_W)c^{z}). \qquad (10.21)
\end{aligned}
$$

Vector boson propagators are given by

$$
\Delta_V^{\mu\nu}(k) = \frac{g^{\mu\nu}}{k^2 - m_V^2 + i\epsilon}; \qquad V = \gamma, W, Z, \qquad (10.22)
$$

with $m_{\gamma} = 0$, while Goldstone boson and ghost propagators are given by

$$
\Delta_s(k^2) = \frac{1}{m_s^2 - k^2 - i\epsilon} \qquad (10.23)
$$

with $m_{G^\pm} = m_{c^\pm} = m_W, m_G = m_{c^z} = m_Z, m_{c^a} = 0$.

The one-loop calculations presented in Chap. 12 will be performed with a modified version of the 't Hooft-Feynman gauge choice, which we may call the e-m gauge invariant 't Hooft-Feynman gauge. The modification amounts to replacing the ordinary derivatives $\partial^\mu W^\pm_\mu$ by electromagnetic covariant derivatives in Eq. (10.13), or equivalently to the replacement

$$f^\pm(\phi_1, \phi_2) \to f^\pm(\phi_1, \phi_2) \mp ieA^\mu W^\pm_\mu. \tag{10.24}$$

As a consequence, the whole Lagrangian, except the free electromagnetic term, becomes invariant under electromagnetic gauge transformations Eqs. (6.17), (6.18) and (6.19).

10.3 Parameters in the Standard Model

The parameters appearing in $\mathcal{L}_{\rm SM}$ are not all independent. The gauge-Higgs sector is entirely specified by the four parameters

$$g, \quad g', \quad v, \quad m_H, \tag{10.25}$$

since

$$m_W^2 = \frac{1}{4}g^2v^2, \quad m_Z^2 = \frac{1}{4}(g^2 + g'^2)v^2, \quad \lambda = \frac{m_H^2}{2v^2}, \quad \tan\theta_W = \frac{g'}{g} \tag{10.26}$$

and $g \sin\theta_W = g' \cos\theta_W = e$. However, v, g, g' are often eliminated in favour of the Fermi constant G_F, the electromagnetic coupling $\alpha_{\rm em}$ and the Z^0 mass m_Z, which are measured with high accuracy. We have

$$G_F = \frac{1}{\sqrt{2}v^2}, \quad \alpha_{\rm em} = \frac{g^2 g'^2}{4\pi(g^2 + g'^2)}, \quad m_Z^2 = \frac{1}{4}(g^2 + g'^2)v^2, \tag{10.27}$$

and therefore

$$v^2 = \frac{1}{\sqrt{2}G_F} \tag{10.28}$$

$$g^2 = 2\sqrt{2}m_Z^2 G_F\left(1 + \sqrt{1 - \frac{4\pi\alpha_{\rm em}}{\sqrt{2}m_Z^2 G_F}}\right) \tag{10.29}$$

$$g'^2 = 2\sqrt{2}m_Z^2 G_F\left(1 - \sqrt{1 - \frac{4\pi\alpha_{\rm em}}{\sqrt{2}m_Z^2 G_F}}\right) \tag{10.30}$$

since $\tan\theta_W < 1$. The free parameters in the fermionic sector are the $3n$ masses $m^f_{\mathrm{U}}, m^f_{\mathrm{D}}, m^f_{\mathrm{E}}$, and the $(n-1)^2$ independent parameters in theCabibbo-Kobayashi-Maskawa matrix V. Including the QCD coupling constant g_s, this gives a total of $4+3n+(n-1)^2+1 = 18$ free parameters for the standard model with three fermion generations.

Chapter 11
Applications

11.1 Muon Decay

The μ^- decays predominantly into an electron, a muon neutrino and an electron anti-neutrino:

$$\mu^-(p) \to e^-(k) + \bar{\nu}_e(k_1) + \nu_\mu(k_2). \tag{11.1}$$

This is a process of great phenomenological importance, since the measurement of the muon lifetime provides the most precise determination of the Fermi constant G_F, and one of the most precise measurements in the whole field of elementary particle physics. In this section, we describe the calculation of the corresponding decay rate in some detail.

The invariant amplitude arises, in the semi-classical approximation, from the single diagram

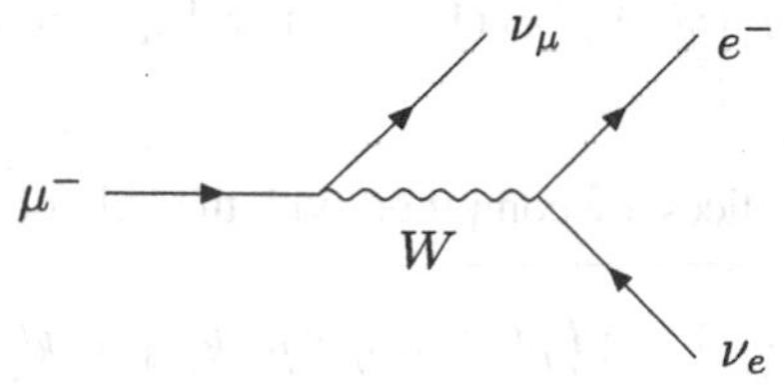

The amplitude is easily computed:

$$\mathcal{M} = \frac{g}{2\sqrt{2}}\,\bar{u}(k_2)\,\gamma^\mu\,(1-\gamma_5)\,u(p)\,\frac{g_{\mu\nu} - \frac{q_\mu q_\nu}{m_W^2}}{q^2 - m_W^2}\,\frac{g}{2\sqrt{2}}\,\bar{u}(k)\,\gamma^\nu\,(1-\gamma_5)\,v(k_1). \tag{11.2}$$

C. M. Becchi and G. Ridolfi, *An Introduction to Relativistic Processes and the Standard Model of Electroweak Interactions*, UNITEXT for Physics,
DOI: 10.1007/978-3-319-06130-6_11, © Springer International Publishing Switzerland 2014

The squared momentum $q^2 = (p - k_2)^2 = (k + k_1)^2$ flowing in the internal W line can be safely neglected with respect to the W mass. This is an excellent approximation, since $q^2 \leq m_\mu^2 \simeq (105.7\ \text{MeV})^2$, while $m_W \simeq 80.4$ GeV. Furthermore, the term $q_\mu q_\nu / m_W^2$ in the numerator of the W propagator gives zero contribution in the limit $m_e = m_\nu = 0$, since

$$(k + k_1)^\mu \, \bar{u}(k) \, \gamma_\mu \, (1 - \gamma_5) \, v(k_1) = \bar{u}(k) \, \not{k} \, (1 - \gamma_5) \, v(k_1) \\ + \bar{u}(k) \, (1 + \gamma_5) \, \not{k}_1 \, v(k_1) = 0 \tag{11.3}$$

by the Dirac equation. Hence, using Eq. (8.3) to relate the constant g to the Fermi constant G_F,

$$\mathcal{M} \simeq -\frac{G_F}{\sqrt{2}} \, \bar{u}(k_2) \, \gamma^\mu \, (1 - \gamma_5) \, u(p) \, \bar{u}(k) \, \gamma_\mu \, (1 - \gamma_5) \, v(k_1). \tag{11.4}$$

The squared amplitude, summed over all spin states of the particles involved, can be computed by reducing spinor products to traces over spinor indices, as we have seen in Sect. 6.2 in the case of the Compton scattering cross section, and using Eqs. (6.62, 6.63) for the sum over fermion spin states. We find

$$\sum_{\text{pol}} |\mathcal{M}|^2 = \frac{G_F^2}{2} \, \text{Tr} \left[\gamma^\mu \, (1 - \gamma_5) \, (\not{p} + m_\mu) \, \gamma^\nu \, (1 - \gamma_5) \, \not{k}_2 \right] \\ \times \text{Tr} \left[\gamma_\mu \, (1 - \gamma_5) \, \not{k}_1 \, \gamma_\nu \, (1 - \gamma_5) \, \not{k} \right]. \tag{11.5}$$

The term proportional to m_μ vanishes, because $(1 + \gamma_5)(1 - \gamma_5) = 0$, and we are left with

$$\sum_{\text{pol}} |\mathcal{M}|^2 = 2 G_F^2 \, \text{Tr} \left[\gamma^\mu \, \not{p} \gamma^\nu \, \not{k}_2 \, (1 + \gamma_5) \right] \text{Tr} \left[\gamma_\mu \, \not{k}_1 \, \gamma_\nu \, \not{k} \, (1 + \gamma_5) \right]. \tag{11.6}$$

The traces over spinor indices are computed with the help of Eqs. (F.8, F.9); we find

$$\text{Tr} \left[\gamma^\mu \, \not{p} \gamma^\nu \, \not{k}_2 \, (1 + \gamma_5) \right] = 4 \left(p^\mu k_2^\nu - g^{\mu\nu} \, p \cdot k_2 + p^\nu k_2^\mu + i \epsilon^{\mu\alpha\nu\beta} \, p_\alpha k_{2\beta} \right), \tag{11.7}$$

and a similar expression for $\text{Tr} \, [\gamma^\mu \, \not{k}_1 \, \gamma^\nu \, \not{k} \, (1 + \gamma_5)]$. The calculation then proceeds in a tedious but straightforward way. The result is remarkably simple:

$$|\mathcal{M}|^2 = 128 \, G_F^2 \, p \cdot k_1 \, k \cdot k_2, \tag{11.8}$$

where we have used the identity

$$\epsilon^{\mu\nu\alpha\beta} \, \epsilon_{\mu\nu\gamma\delta} = -2 \left(\delta_\gamma^\alpha \, \delta_\delta^\beta - \delta_\delta^\alpha \, \delta_\gamma^\beta \right). \tag{11.9}$$

The differential decay rate is given by Eq. (4.38):

$$d\Gamma = \frac{1}{2m_\mu}\frac{1}{2}\sum_{\text{pol}}|\mathcal{M}|^2\, d\phi_3(p;k,k_1,k_2), \tag{11.10}$$

where we have averaged over the two spin states of the decaying muon, and

$$d\phi_3(p;k,k_1,k_2) = \frac{d^3k}{(2\pi)^3 2k^0}\frac{d^3k_1}{(2\pi)^3 2k_1^0}\frac{d^3k_2}{(2\pi)^3 2k_2^0}(2\pi)^4\delta^{(4)}(p-k-k_1-k_2). \tag{11.11}$$

An interesting quantity from the phenomenological point of view is the electron energy spectrum $d\Gamma/dE$. This can be computed by integrating Eq. (11.10) over neutrino momenta. Recalling the expression Eq. (11.8) of the squared amplitude, we see that in order to obtain the electron energy spectrum we must compute

$$I^{\alpha\beta}(K) = \int \frac{d^3k_1}{k_1^0}\frac{d^3k_2}{k_2^0}\, k_1^\alpha\, k_2^\beta\, \delta^{(4)}(K-k_1-k_2), \tag{11.12}$$

where we have defined $K = p - k = k_1 + k_2$. Because the integration measure is Lorentz invariant, $I^{\alpha\beta}(K)$ is a tensor, and it only depends on the four-momentum K. Hence, it has necessarily the form

$$I^{\alpha\beta}(K) = A\, K^\alpha K^\beta + B\, K^2\, g^{\alpha\beta}. \tag{11.13}$$

The constants A and B can be determined as follows. We observe that, for massless neutrinos,

$$k_1\cdot k_2 = \frac{1}{2}(k_1+k_2)^2 = \frac{K^2}{2}; \qquad K\cdot k_1\, K\cdot k_2 = (k_1\cdot k_2)^2 = \frac{K^4}{4} \tag{11.14}$$

and we multiply Eq. (11.13) by $g_{\alpha\beta}$ and $K_\alpha K_\beta/K^2$. We obtain the system of equations

$$A + 4B = \frac{I}{2} \tag{11.15}$$

$$A + B = \frac{I}{4} \tag{11.16}$$

where

$$I = \int \frac{d^3k_1}{k_1^0}\frac{d^3k_2}{k_2^0}\,\delta^{(4)}(K-k_1-k_2) = \int \frac{d^3k_1}{|\boldsymbol{k}_1|^2}\,\delta(K^0 - 2\,|\boldsymbol{k}_1|) = 2\pi, \tag{11.17}$$

and therefore

$$A = \frac{\pi}{3}; \qquad B = \frac{\pi}{6}. \tag{11.18}$$

Replacing this result in Eq. (11.13), and then in Eq. (11.10), we find

$$d\Gamma = \frac{G_F^2}{12\pi^3}\,(3m_\mu^2 - 4m_\mu E)\,E^2\,dE, \tag{11.19}$$

where E is the energy of the emitted electron in the rest frame of the decaying muon.

The total rate is now easily computed. The maximum value of E is achieved when the two neutrinos are emitted in the same direction; in this case, $E = E_{\max} = m_\mu/2$. Hence,

$$\Gamma = \frac{G_F^2}{12\pi^3}\int_0^{m_\mu/2} dE\,E^2\,(3m_\mu^2 - 4m_\mu E) = \frac{G_F^2 m_\mu^5}{192\pi^3}. \tag{11.20}$$

11.2 The Decay Rate of the W Boson

A simple and interesting application of what we have learned about relativistic processes and the standard model is the computation of the decay rate of the W boson into a quark–antiquark pair:

$$W^+(p,\epsilon) \to u_f(k_1) + \bar{d}_g(k_2), \tag{11.21}$$

where f, g are generation indices, and particle four-momenta are displayed in brackets. In the semi-classical approximation, the only relevant diagram is

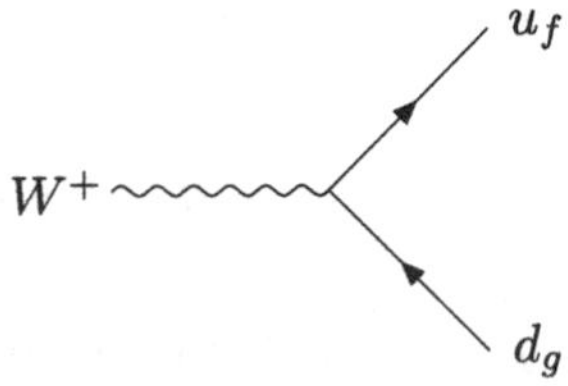

that corresponds to the invariant amplitude

$$\mathcal{M} = \frac{g}{2\sqrt{2}}\,V_{fg}\,\bar{u}(k_1)\,\gamma^\mu\,(1-\gamma_5)\,v(k_2)\,\epsilon_\mu(p), \tag{11.22}$$

where ϵ is the polarization vector of the W, and $p = k_1 + k_2$. We get

$$\sum_{\text{pol}} |\mathcal{M}|^2 = \frac{3g^2\,|V_{fg}|^2}{4}\,\text{Tr}\left[\gamma^\mu\,(\not{k}_2 - m_d)\,\gamma^\nu\,(1-\gamma_5)\,(\not{k}_1 + m_u)\right]\left(-g_{\mu\nu} + \frac{p_\mu p_\nu}{m_W^2}\right), \tag{11.23}$$

where we have summed over the three polarization states of the W boson using Eq. (8.46), and we have inserted a factor of 3 that accounts for the three different colour states of the quark–antiquark pair. Dropping terms proportional to an odd number of γ matrices, and using $(1+\gamma_5)(1-\gamma_5)=0$, we obtain

$$\sum_{\text{pol}} |\mathcal{M}|^2 = \frac{3g^2}{4} \left|V_{fg}\right|^2 \text{Tr}\left[\gamma^\mu \not{k}_2 \gamma^\nu \not{k}_1\right] \left(-g_{\mu\nu} + \frac{p_\mu p_\nu}{m_W^2}\right). \tag{11.24}$$

The calculation is now straightforward; using the formulae obtained in Appendix F for the trace of products of two and four γ matrices we obtain

$$\sum_{\text{pol}} |\mathcal{M}|^2 = 3g^2 \left|V_{fg}\right|^2 \left(k_1 \cdot k_2 + \frac{2p \cdot k_1\, p \cdot k_2}{m_W^2}\right). \tag{11.25}$$

The values of the scalar products can be obtained by taking the square of the momentum conservation relation, and using the mass-shell conditions $p^2 = m_W^2, k_1^2 = m_u^2, k_2^2 = m_d^2$. We find

$$k_1 \cdot k_2 = \frac{m_W^2 - m_u^2 - m_d^2}{2} \tag{11.26}$$

$$p \cdot k_1 = \frac{m_W^2 + m_u^2 - m_d^2}{2} \tag{11.27}$$

$$p \cdot k_2 = \frac{m_W^2 - m_u^2 + m_d^2}{2}. \tag{11.28}$$

The kinematically-allowed channels for W decay into a quark–antiquark pair include all combinations $u_f\, \bar{d}_g$, except those containing the top quark, whose mass is around 173 GeV, more than twice the W mass. All other quark masses are much smaller than m_W, the largest of them being $m_b \simeq 5$ GeV. We may therefore neglect fermion masses in Eq. (11.25) to an excellent degree of accuracy. In this limit,

$$k_1 \cdot k_2 \simeq p \cdot k_1 \simeq p \cdot k_2 \simeq \frac{m_W^2}{2}, \tag{11.29}$$

and we find

$$\sum_{\text{pol}} |\mathcal{M}|^2 \simeq 3g^2 m_W^2 \left|V_{fg}\right|^2. \tag{11.30}$$

The differential decay rate is now immediately computed using Eq. (4.38) and the expression Eq. (4.25) for the two-body phase space. After averaging Eq. (11.30) over the three polarization states of the W boson, we get

$$d\Gamma(W \to u_f \bar{d}_g) = \frac{3g^2 m_W \left|V_{fg}\right|^2}{96\pi} d\cos\theta = \frac{3G_F m_W^3 \left|V_{fg}\right|^2}{12\sqrt{2}\pi} d\cos\theta, \quad (11.31)$$

where θ is the angle formed by the direction of the decay products and the Z axis in the rest frame of the decaying W.

The total decay rate of the W boson is obtained by performing the (trivial) angular integration, summing over the allowed quark–antiquark channels, and adding the contributions of the three leptonic channels, given by Eq. (11.31) with the replacement $3\left|V_{fg}\right|^2 \to 1$. The final result is

$$\Gamma_W = \frac{G_F m_W^3}{6\sqrt{2}\pi}\left[3\sum_{f=u,c}\sum_{g=d,s,b}\left|V_{fg}\right|^2 + 3\right]. \quad (11.32)$$

Neglecting, to a first approximation, the mixing of the third generation, we have

$$\sum_{f=u,c}\sum_{g=d,s,b}\left|V_{fg}\right|^2 \simeq \sum_{f=u,c}\sum_{g=d,s}\left|V_{fg}\right|^2 \simeq 2(\cos^2\theta_c + \sin^2\theta_c) = 2, \quad (11.33)$$

and therefore

$$\Gamma_W = \frac{3G_F m_W^3}{2\sqrt{2}\pi} \simeq 2.05\ \text{GeV}, \quad (11.34)$$

in good agreement with the measured value

$$\Gamma_W^{\text{exp}} = 2.085 \pm 0.042\ \text{GeV}. \quad (11.35)$$

11.3 Higgs Decay into a Vector Boson Pair

In this section, we consider the decay of a Higgs boson into a W^+W^- pair:

$$H(p) \to W^+(k_1, \epsilon_1) + W^-(k_2, \epsilon_2). \quad (11.36)$$

This decay channel is now known to be kinematically forbidden to the standard model Higgs boson, whose mass is much less than twice the W mass. It is however interesting to study it in some detail, because heavy scalar states typically appear in many extensions of the standard model.

In the semiclassical approximation, the invariant amplitude is given by the single diagram

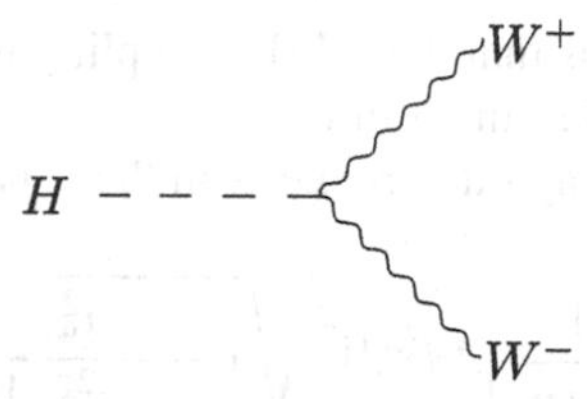

and from Eq. (10.10) we obtain

$$\mathcal{M} = \frac{2m_W^2}{v} \epsilon_1^* \cdot \epsilon_2^*. \tag{11.37}$$

Observe that the coupling of the Higgs boson to W bosons is proportional to the W mass, which is a general feature of Higgs couplings.

In a reference frame in which the decaying Higgs is at rest, the two W bosons are emitted with three-momenta $\pm\boldsymbol{k}$, in the same direction:

$$p = (m_H, \boldsymbol{0}); \qquad k_1 = \left(\frac{m_H}{2}, \boldsymbol{k}\right); \qquad k_2 = \left(\frac{m_H}{2}, -\boldsymbol{k}\right) \tag{11.38}$$

with

$$|\boldsymbol{k}| = \frac{m_H}{2}\sqrt{1 - \frac{4m_W^2}{m_H^2}}. \tag{11.39}$$

Since the Higgs boson is a spin-0 particle, the amplitude is invariant under rotations, and we may choose to orientate the z axis in the direction of the three-momenta of the decay products. The polarization vectors ϵ_1, ϵ_2 in the rest frame of each of the W bosons are given in Eq. (8.36). After a Lorentz boost to the Higgs rest frame, $\epsilon^{(1)}$ and $\epsilon^{(2)}$ are unchanged, while

$$\epsilon^{(3)}(k_1) = \frac{1}{m_W}\begin{pmatrix} -|\boldsymbol{k}| \\ 0 \\ 0 \\ m_H/2 \end{pmatrix}; \qquad \epsilon^{(3)}(k_2) = \frac{1}{m_W}\begin{pmatrix} |\boldsymbol{k}| \\ 0 \\ 0 \\ m_H/2 \end{pmatrix}. \tag{11.40}$$

It follows that the amplitude $\mathcal{M}$ is nonzero only when the two polarization vectors are parallel. The corresponding amplitudes are

$$\mathcal{M}^{(11)} = \mathcal{M}^{(22)} = -\frac{2m_W^2}{v} \tag{11.41}$$

$$\mathcal{M}^{(33)} = -\frac{2m_W^2}{v}\frac{1}{m_W^2}\left(|\boldsymbol{k}|^2 + \frac{m_H^2}{4}\right) = -\frac{1}{v}\left(m_H^2 - 2m_W^2\right). \tag{11.42}$$

It is interesting to observe that $\mathcal{M}^{(11)}$ and $\mathcal{M}^{(22)}$ vanish for $m_W \to 0$ with v fixed (or equivalently in the limit $g \to 0$, in which gauge interactions are switched off), as one would expect on the basis that the HWW coupling is proportional to m_W, while $\mathcal{M}^{(33)}$ remains nonzero in the same limit.

The calculation of the decay rate proceeds in the usual way; we find

$$d\Gamma^{(i)} = \frac{1}{2m_H} \left|\mathcal{M}^{(ii)}\right|^2 \sqrt{1 - \frac{4m_W^2}{m_H^2}} \frac{1}{16\pi} d\cos\theta \tag{11.43}$$

and

$$\begin{aligned} \Gamma(H \to WW) &= \int \sum_{i=1}^{3} d\Gamma^{(i)} \\ &= \frac{G_F m_H^3}{8\sqrt{2}\pi} \sqrt{1 - \frac{4m_W^2}{m_H^2}} \left[1 + \frac{12m_W^4}{m_H^4} - \frac{4m_W^2}{m_H^2}\right], \end{aligned} \tag{11.44}$$

where we have used $G_F = \frac{1}{\sqrt{2}v^2}$, Eq. (10.27).

We would have obtained exactly the same result, up to a factor 1/2 due to the identity of final particles, if we had computed the decay rate $\Gamma(H \to \gamma\gamma)$ of the scalar H into a pair of massive photons in the context of the abelian Higgs model of Sect. 8.2. One can check that, in the limit $m_\gamma = ev \to 0$ with v fixed, $\Gamma(H \to \gamma\gamma)$ is equal to the decay rate $\Gamma(H \to GG)$ of the scalar H into a pair of Goldstone bosons. This is a particular case of a general result, sometimes called the *equivalence theorem*: at energies much larger than the vector boson mass, the couplings of the longitudinal components of vector bosons are the same as those of the corresponding Goldstone scalars (recall that the decay rate into pairs of transversally polarized W's vanishes as $m_W \to 0$).

11.4 Weak Neutral Currents

The first experimental confirmation of the standard model was the observation of the effects of weak neutral currents, that allowed a measurement of $\sin\theta_W$ and therefore an estimate of the masses of the intermediate vector bosons W and Z. The relevant process is inclusive neutrino scattering off nuclear targets,

$$\text{NC}: \quad \nu_\mu(p_1) + N(p) \to \nu_\mu(k_1) + X \tag{11.45}$$

$$\text{CC}: \quad \nu_\mu(p_1) + N(p) \to \mu^-(k_1) + X, \tag{11.46}$$

where N is a nuclear target with the same number of protons and neutrons and X any hadronic final state. In the parton model, the cross section for a process with a hadron H of momentum p in the initial state is given by

$$\sigma(p) = \sum_q \int_0^1 dz\, f_q^H(z)\, \sigma_q(zp), \tag{11.47}$$

where q is a generic parton, $\sigma_q(zp)$ is the cross section for the same process with parton q with momentum zp in the initial state, and $f_q^H(z)$ are universal functions, that characterize the structure of the hadron H: $f_q^H(z)\,dz$ is the probability that parton q carries a fraction of the hadron momentum between z and $z+dz$. The cross section for the process Eq. (11.45) is therefore given by the incoherent sum of contributions from proton and neutron cross sections:

$$\begin{aligned}\sigma_\nu^{(NC)}(p) &= \frac{1}{2}\left[\sigma_{\nu P}^{(NC)}(p) + \sigma_{\nu N}^{(NC)}(p)\right] \\ &= \int_0^1 dz \left[\frac{f_u^P(z) + f_u^N(z)}{2}\,\sigma_{\nu u}^{(NC)}(zp) + \frac{f_d^P(z) + f_d^N(z)}{2}\,\sigma_{\nu d}^{(NC)}(zp)\right] \\ &= \int_0^1 dz\, \frac{f_u(z) + f_d(z)}{2}\left[\sigma_{\nu u}^{(NC)}(zp) + \sigma_{\nu d}^{(NC)}(zp)\right],\end{aligned} \tag{11.48}$$

where nuclear charge symmetry implies $f_{u,d}(z) = f_{u,d}^P(z) = f_{d,u}^N(z)$. A similar expression is found for $\sigma_\nu^{(CC)}(p)$.

In the case of the processes in Eqs. (11.45, 11.46), the relevant parton subprocesses are

$$\text{NC}: \quad \nu_\mu(p_1) + q(p_2) \to \nu_\mu(k_1) + q(k_2); \qquad q = u, d \tag{11.49}$$

$$\text{CC}: \quad \nu_\mu(p_1) + d(p_2) \to \mu^-(k_1) + u(k_2) \tag{11.50}$$

The relevant lowest-order diagrams are

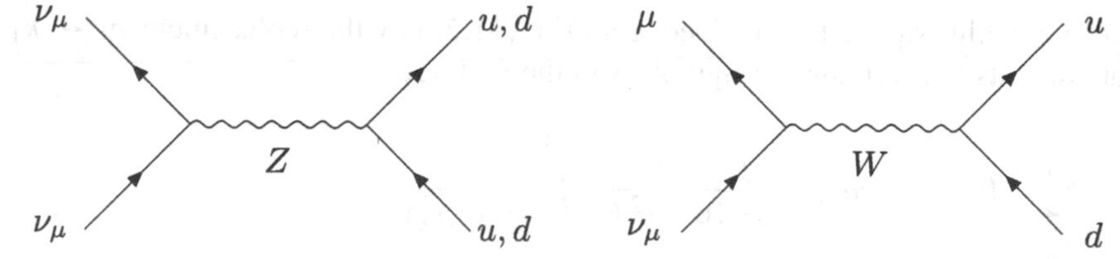

and the relevant Feynman rules are read off the following terms of the standard model Lagrangian:

$$\mathcal{L} = \frac{g}{4\cos\theta_W}\left[\bar{\nu}_\mu\, \gamma_\mu(1-\gamma_5)\, \nu_\mu + \bar{u}\, \gamma_\mu\, (V_u - A_u\gamma_5)\, u + \bar{d}\, \gamma_\mu\, (V_d - A_d\gamma_5)\, d\right] Z^\mu$$

$$+\frac{g}{2\sqrt{2}}\left[\bar{\nu}_\mu\,\gamma^\mu(1-\gamma_5)\,e+\bar{u}\,\gamma^\mu(1-\gamma_5)\,V_{ud}\,d\right]W^+_\mu$$
$$+\frac{g}{2\sqrt{2}}\left[\bar{e}\,\gamma^\mu(1-\gamma_5)\,\nu_\mu+\bar{d}\,\gamma^\mu(1-\gamma_5)\,V^*_{ud}\,u\right]W^-_\mu, \qquad (11.51)$$

where

$$V_u = 1-\frac{8}{3}\sin^2\theta_W; \qquad A_u = 1. \qquad (11.52)$$
$$V_d = -1+\frac{4}{3}\sin^2\theta_W; \qquad A_d = -1. \qquad (11.53)$$

Neglecting all fermion masses we find

$$\mathcal{M}(\nu q\to\nu q) = \frac{g^2}{16\cos^2\theta_W}\frac{1}{t-m_Z^2}$$
$$\bar{u}(k_1)\gamma^\mu(1-\gamma_5)u(p_1)\,\bar{u}(k_2)\gamma_\mu(V_q-A_q\gamma_5)u(p_2), \quad (11.54)$$

where $t=-2p_1\cdot k_1$, and therefore

$$\sum_{\text{pol}}|\mathcal{M}(\nu q\to\nu q)|^2 = \left(\frac{g^2}{16\cos^2\theta_W}\right)^2\frac{1}{(t-m_Z^2)^2}$$
$$2\mathrm{Tr}\left[\gamma^\mu \not{p}_1\gamma^\nu \not{k}_1(1+\gamma_5)\right]\,\mathrm{Tr}\left[\gamma_\mu \not{p}_2\gamma_\nu \not{k}_2(V_q^2+A_q^2+2V_qA_q\gamma_5)\right]. \quad (11.55)$$

The amplitude for the conjugated process is given by

$$\mathcal{M}(\bar{\nu} q\to\bar{\nu} q) = \frac{g^2}{16\cos^2\theta_W}\frac{1}{t-m_Z^2}$$
$$\bar{v}(p_1)\gamma^\mu(1-\gamma_5)v(k_1)\,\bar{u}(k_2)\gamma_\mu(V_q-A_q\gamma_5)u(p_2), \quad (11.56)$$

whose modulus squared is obtained from Eq. (11.55) by the replacement $p_1\leftrightarrow k_1$, that amounts to reversing the sign of γ_5 in the first trace:

$$\sum_{\text{pol}}|\mathcal{M}(\bar{\nu} q\to\bar{\nu} q)|^2 = \left(\frac{g^2}{16\cos^2\theta_W}\right)^2\frac{1}{(t-m_Z^2)^2}$$
$$2\mathrm{Tr}\left[\gamma^\mu \not{p}_1\gamma^\nu \not{k}_1(1-\gamma_5)\right]\,\mathrm{Tr}\left[\gamma_\mu \not{p}_2\gamma_\nu \not{k}_2(V_q^2+A_q^2+2V_qA_q\gamma_5)\right]. \quad (11.57)$$

An analogous calculation yields the result

$$\sum_{\text{pol}} |\mathcal{M}(\nu d \to \mu u)|^2 = \left(\frac{g^2\,|V_{ud}|}{8}\right)^2 \frac{1}{(t-m_W^2)^2}$$

$$4\text{Tr}\left[\gamma^\mu \not{p}_1 \gamma^\nu \not{k}_1 (1+\gamma_5)\right] \text{Tr}\left[\gamma_\mu \not{p}_2 \gamma_\nu \not{k}_2 (1+\gamma_5)\right]. \quad (11.58)$$

$$\sum_{\text{pol}} |\mathcal{M}(\bar{\nu} u \to \mu d)|^2 = \left(\frac{g^2\,|V_{ud}|}{8}\right)^2 \frac{1}{(t-m_W^2)^2}$$

$$4\text{Tr}\left[\gamma^\mu \not{p}_1 \gamma^\nu \not{k}_1 (1-\gamma_5)\right] \text{Tr}\left[\gamma_\mu \not{p}_2 \gamma_\nu \not{k}_2 (1+\gamma_5)\right]. \quad (11.59)$$

The parton cross sections are given by the general formula Eq. (4.34); in our cases,

$$\sigma = \frac{1}{4s} \int d\phi_2(p_1 + p_2; k_1, k_2)\ |\mathcal{M}|^2, \quad (11.60)$$

where $s = 2p_1 \cdot p_2$, we have inserted a factor of 1/2 to average over initial spin states, and we have assumed fermions to be massless. It is easy to show that the differences

$$\sum_{\text{pol}} \left[|\mathcal{M}(\bar{\nu} q \to \bar{\nu} q)|^2 - |\mathcal{M}(\nu q \to \nu q)|^2\right] \quad (11.61)$$

$$\sum_{\text{pol}} \left[|\mathcal{M}(\bar{\nu} u \to \mu d)|^2 - |\mathcal{M}(\nu u \to \mu d)|^2\right] \quad (11.62)$$

are linear in the initial parton momentum p_2; since s is also linear in p_2, it follows that the parton cross section differences are independent of the momentum fraction z carried by the initial parton, $p_2 = zp$. Thus, using Eq. (11.48) we find

$$\sigma_{\bar{\nu}}^{(\text{NC})} - \sigma_{\nu}^{(\text{NC})} = \int_0^1 dz\, \frac{f_u(z) + f_d(z)}{2} \left[\sigma_{\bar{\nu}u}^{(\text{NC})} - \sigma_{\nu u}^{(\text{NC})} + \sigma_{\bar{\nu}d}^{(\text{NC})} - \sigma_{\nu d}^{(\text{NC})}\right]$$

$$= \int_0^1 dz\, \frac{f_u(z) + f_d(z)}{2} \frac{2}{S} \left(\frac{g^2}{16 m_Z^2 \cos^2\theta_W}\right)^2 (-V_u A_u - V_d A_d)$$

$$\times \int d\phi_2\, \text{Tr}\left[\gamma^\mu \not{p}_1 \gamma^\nu \not{k}_1 \gamma_5\right] \text{Tr}\left[\gamma_\mu \not{p} \gamma_\nu \not{k}_2 \gamma_5\right] \quad (11.63)$$

where $S = 2p_1 \cdot p$, and similarly

$$\sigma_{\bar{\nu}}^{(\text{CC})} - \sigma_{\nu}^{(\text{CC})} = \int_0^1 dz\, \frac{f_u(z) + f_d(z)}{2} \left[\sigma_{\bar{\nu}u}^{(\text{CC})} - \sigma_{\nu d}^{(\text{CC})}\right]$$

$$= \int_0^1 dz \, \frac{f_u(z) + f_d(z)}{2} \frac{2}{S} \left(\frac{g^2 \, |V_{ud}|}{8m_W^2} \right)^2$$

$$\times \int d\phi_2 \, \mathrm{Tr} \left[\gamma^\mu \not{p}_1 \gamma^\nu \not{k}_1 \gamma_5 \right] \mathrm{Tr} \left[\gamma_\mu \not{p} \gamma_\nu \not{k}_2 \gamma_5 \right]. \quad (11.64)$$

Most factors, including those that depend on the parton distribution functions, cancel in the ratio

$$R = \frac{\sigma_{\bar{\nu}}^{(\mathrm{NC})} - \sigma_\nu^{(\mathrm{NC})}}{\sigma_{\bar{\nu}}^{(\mathrm{CC})} - \sigma_\nu^{(\mathrm{CC})}}, \quad (11.65)$$

which therefore provides a direct measure of $\sin \theta_W$. We find

$$R = -\frac{V_u A_u + V_d A_d}{4 \, |V_{ud}|^2} = \frac{1 - 2 \sin^2 \theta_W}{2 \, |V_{ud}|^2}, \quad (11.66)$$

which is the so-called Paschos-Wolfenstein relation.

11.5 Higgs Production in e^+e^- Collisions

The Higgs boson can be produced in electron-positron collisions in association with a Z^0 boson:

$$e^-(p_1) + e^+(p_2) \to H(k_1) + Z^0(k_2, \epsilon). \quad (11.67)$$

The relevant diagram in the semi-classical approximation is

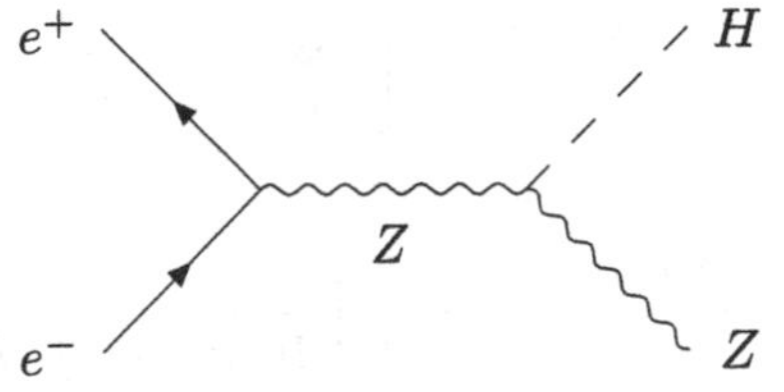

The relevant terms in the interaction Lagrangian are the weak neutral-current coupling of the electron with the Z boson, and the HZZ coupling, which we rewrite here:

$$\mathcal{L} = \frac{g}{4 \cos \theta_W} \bar{e} \, \gamma_\mu \left(-1 + 4 \sin^2 \theta_W + \gamma_5 \right) e \, Z^\mu + \frac{m_Z^2}{v} \, H \, Z^\mu Z_\mu. \quad (11.68)$$

This gives an invariant amplitude

$$\mathcal{M} = \frac{g}{\cos\theta_W}\frac{m_Z^2}{2v}\,\epsilon^{*\mu}(k_2)\,\frac{g_{\mu\nu} - \frac{q_\mu q_\nu}{m_Z^2}}{q^2 - m_Z^2}\,\bar{v}(p_2)\,\gamma^\nu\,(V - A\gamma_5)\,u(p_1), \tag{11.69}$$

where $q = p_1 + p_2$ and

$$V = -1 + 4\sin^2\theta_W; \qquad A = -1. \tag{11.70}$$

It will be convenient to express the amplitude in terms of the Fermi constant G_F; to this purpose, we observe that

$$\frac{g}{\cos\theta_W}\frac{m_Z}{2v} = \sqrt{2}\,G_F\,m_Z^2. \tag{11.71}$$

Neglecting the electron mass, which is perfectly adequate in this energy regime, we can ignore the $q_\mu q_\nu$ term in the internal propagator, and compute the unpolarized squared amplitude in the usual way:

$$\sum_{\text{pol}} |\mathcal{M}|^2 = \frac{2\,G_F^2\,m_Z^6}{(s - m_Z^2)^2}\left(-g^{\mu\nu} + \frac{k_2^\mu k_2^\nu}{m_Z^2}\right)\text{Tr}\,\left[\gamma_\mu\,(V - A\gamma_5)\,\not{p}_1\,\gamma_\nu\,(V - A\gamma_5)\,\not{p}_2\right] \tag{11.72}$$

where $s = (p_1 + p_2)^2$. Terms proportional to γ_5 are proportional to the product of an antisymmetric ϵ tensor with the sum over Z polarization, which is symmetric; hence, they do not contribute to the unpolarized cross section, and Eq. (11.72) becomes

$$\begin{aligned}\sum_{\text{pol}} |\mathcal{M}|^2 &= \frac{2\,G_F^2\,m_Z^6\,(V^2 + A^2)}{(s - m_Z^2)^2}\left(-g^{\mu\nu} + \frac{k_2^\mu k_2^\nu}{m_Z^2}\right)\text{Tr}\,\left[\gamma_\mu\,\not{p}_1\,\gamma_\nu\,\not{p}_2\right] \\ &= \frac{4\,G_F^2\,m_Z^6\,(V^2 + A^2)}{(s - m_Z^2)^2}\left[s + \frac{(t - m_Z^2)(u - m_Z^2)}{m_Z^2}\right],\end{aligned} \tag{11.73}$$

where we have used Eq. (F.8) and we have defined

$$t = (p_1 - k_2)^2; \qquad u = (p_2 - k_2)^2. \tag{11.74}$$

The invariants s, t, u are related by

$$s + t + u = m_Z^2 + m_H^2 \tag{11.75}$$

through energy-momentum conservation.

In the center-of-mass frame of the colliding leptons we have

$$t = m_Z^2 - \sqrt{s}\left(E_{\boldsymbol{k}_2} - |\boldsymbol{k}_2|\cos\theta\right) \tag{11.76}$$

$$u = m_Z^2 - \sqrt{s}\left(E_{\boldsymbol{k}_2} + |\boldsymbol{k}_2|\cos\theta\right), \tag{11.77}$$

where θ is the scattering angle of the Z boson in the final state, and the value of $|\boldsymbol{k}_2|$ is found by solving the energy conservation constraint

$$\sqrt{s} = \sqrt{|\boldsymbol{k}_2|^2 + m_Z^2} + \sqrt{|\boldsymbol{k}_2|^2 + m_H^2}; \tag{11.78}$$

we find

$$\begin{aligned} |\boldsymbol{k}_2|^2 &= \frac{\lambda(s, m_Z^2, m_H^2)}{4s}; \quad \lambda(x,y,z) = x^2 + y^2 + z^2 - 2xy - 2xz - 2yz \\ E_{\boldsymbol{k}_2} &= \frac{s + m_Z^2 - m_H^2}{2\sqrt{s}}. \end{aligned} \tag{11.79}$$

Thus,

$$\sum_{\text{pol}} |\mathcal{M}|^2 = \frac{4\,G_F^2\, m_Z^4\,(V^2 + A^2)\, s}{(s - m_Z^2)^2}\left(m_Z^2 + E_{\boldsymbol{k}_2}^2 - |\boldsymbol{k}_2|^2\cos^2\theta\right). \tag{11.80}$$

In terms of center-of-mass quantities, the two body phase-space measure is given by

$$d\phi_2(p_1 + p_2; k_1, k_2) = \frac{1}{8\pi}\frac{|\boldsymbol{k}_2|}{\sqrt{s}}\, d\cos\theta. \tag{11.81}$$

The differential cross section is therefore given by

$$d\sigma = \frac{G_F^2\, m_Z^4\,(V^2 + A^2)}{16\pi}\frac{|\boldsymbol{k}_2|}{\sqrt{s}}\frac{m_Z^2 + E_{\boldsymbol{k}_2}^2 - |\boldsymbol{k}_2|^2\cos^2\theta}{(s - m_Z^2)^2}\, d\cos\theta \tag{11.82}$$

and the total cross section by

$$\sigma(e^+e^- \to HZ) = \frac{G_F^2\, m_Z^4\,(V^2 + A^2)}{12\pi}\frac{|\boldsymbol{k}_2|}{\sqrt{s}}\frac{3m_Z^2 + |\boldsymbol{k}_2|^2}{(s - m_Z^2)^2}. \tag{11.83}$$

Chapter 12
Beyond the Classical Approximation

12.1 The General Structure of Loop Diagrams and the Corresponding Amplitudes

We have seen in Chap. 4 that physical transition amplitudes correspond to amputated connected diagrams with a given number of external lines. In general, these diagrams contain loops as discussed in Sect. 4.5. Following the Feynman rules given in Chap. 4, a generic connected diagram can be built as a tree diagram whose vertices are either point-like, or sub-diagrams with loops. We shall call these vertices *effective vertices*: an effective vertex is a sub-diagram that cannot be broken into two parts cutting a single internal line. For this reason, they are also called *irreducible* diagrams. As an example, consider the diagram

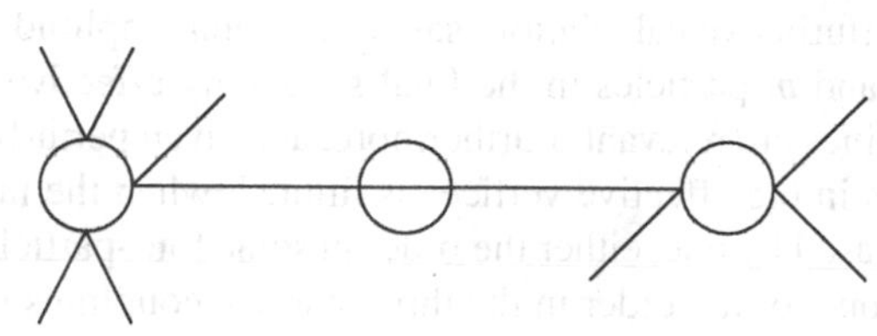

which has three effective vertices with six, two and four external lines. Its tree (or *skeleton*) structure is

C. M. Becchi and G. Ridolfi, *An Introduction to Relativistic Processes and the Standard Model of Electroweak Interactions*, UNITEXT for Physics,
DOI: 10.1007/978-3-319-06130-6_12,

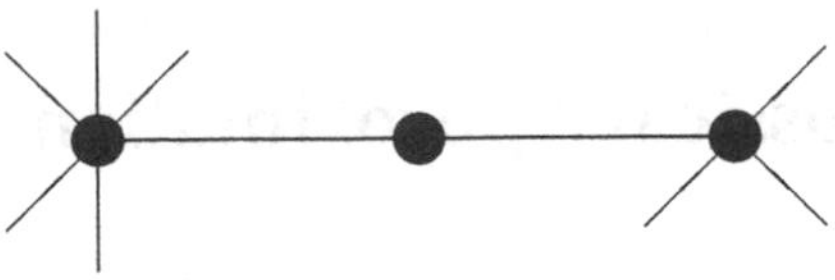

where black dots denote effective vertices.

Once the amplitudes corresponding to irreducible diagrams are known, one can build an *effective action* as a generalization of the classical action in which the coefficient of the individual terms, e.g. the coupling constants, are replaced by the sum of all effective vertices with the same external lines, starting from the classical vertices.[1] For example, in the scalar field theory of Eqs. (2.5, 2.10) the ϕ^4 term in the effective action has the structure

$$S^{\phi^4}_{\text{eff}} = \frac{1}{4!}\int \prod_{i=1}^{4}\left[\hat{\phi}(p_i)dp_i\right](2\pi)^4\delta(p_1+p_2+p_3+p_4)\gamma_4(p_1,\ldots,p_4), \quad (12.1)$$

where $\hat{\phi}(p)$ is the Fourier transform of the field, defined as in Eq. (4.7), and

$$\gamma_4(p_1,\ldots,p_4) = -\lambda + 3\mathcal{M}((p_1+p_2)^2) + O(\lambda^3), \quad (12.2)$$

where $\mathcal{M}$ is given in Eq. (4.64). It follows that the effective action is the sum of an infinite number of terms, each of which is the sum of the contributions of an infinite number of irreducible diagrams. Connected amplitudes correspond to sums of tree diagrams generated by the effective action in which the effective vertices are the sum of the irreducible diagrams with the prescribed number of external lines.

Computing the full effective action is obviously an impossible task. However, in a practical perturbative calculation, say a transition amplitude with n particles in the initial state and m particles in the final state, only effective vertices with up to $n+m$ external lines are relevant. Furthermore, at a given perturbative order the total number of loops in the effective vertices is limited: when the number of loops in a diagram is increased by one, either the order in some four-particle coupling constant is increased by one, or the order in the three-particle couplings is increased by two.

12.2 The Decay $H \to \gamma\gamma$

An interesting example of a calculation beyond the classical approximation is that of the decay width of a Higgs boson into two photons. There is no vertex in the classical Lagrangian which couples a Higgs to two photon lines, and hence the decay amplitude vanishes in the tree approximation, and the process takes place at the level of loop

[1] Here we call vertices also the terms of the free classical action which are bilinear in the fields.

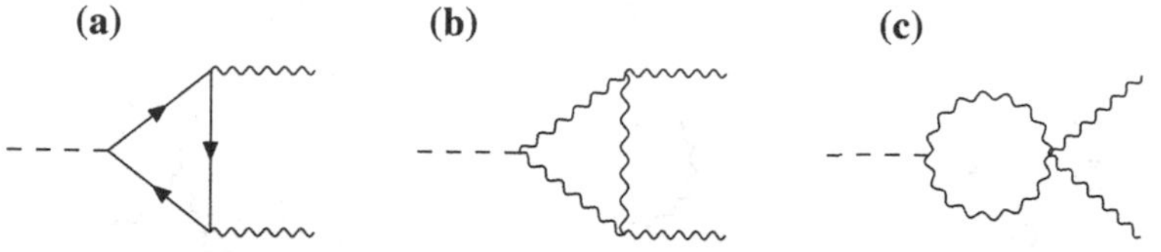

Fig. 12.1 One-loop diagrams for $H \to \gamma\gamma$

diagrams. At one loop, these diagrams contain either a charged, massive fermion or a charged boson closed line as shown in Fig. 12.1. Diagram (a) represents a matter fermion loop, while diagrams (b) and (c) are W loop diagrams.

Taking into account Lorentz and gauge invariance one sees that the effective vertices responsible for $H \to \gamma\gamma$ correspond to a term

$$\begin{aligned} S^{H\gamma\gamma}_{\rm eff} &= -\frac{1}{4}\int d^4x\, d^4y\, d^4z\, \gamma^{\mu\nu\rho\sigma}(y-x, z-x) H(x) F_{\mu\nu}(y) F_{\rho\sigma}(z) \\ &= -\frac{1}{4}\int d^4k\, d^4p\, d^4q\, \hat{H}(k)\hat{F}_{\mu\nu}(p)\hat{F}_{\rho\sigma}(q)\hat{\gamma}^{\mu\nu\rho\sigma}(k,p,q)(2\pi)^4\delta(k+p+q) \end{aligned} \tag{12.3}$$

in the effective action. This expression simplifies considerably when the particle momenta are small with respect to the intermediate particle masses:

$$S^{H\gamma\gamma}_{\rm eff} = -\frac{1}{4}A\int d^4x\, H(x) F^{\mu\nu}(x) F_{\mu\nu}(x) \tag{12.4}$$

where A is a constant coefficient, and we have excluded a term proportional to $\epsilon_{\mu\nu\rho\sigma}F^{\mu\nu}(x)F^{\rho\sigma}(x)$ on the basis of CP invariance of quark and W couplings to the Higgs and of electromagnetic interactions.

The diagrams shown in Fig. 12.1 are not easy to compute. In particular, W propagators and couplings depend on the adopted gauge-fixing procedure (see Sect. 10.2) and in general involve in a non-trivial way the Goldstone bosons introduced in Sect. 8.2. However, we shall see that a good approximation to the exact amplitude is obtained in the limit $\lambda \to 0$ in Eq. (8.69). In this limit, the Higgs mass vanishes (see Eq. 8.72). As a consequence, the $H \to \gamma\gamma$ physical transition amplitude is computed in the limit of vanishing particle momenta, and is therefore proportional to the constant A.

The important fact is that in the limit $\lambda \to 0$ the classical theory specified by the action built in Chap. 8 acquires a *dilation symmetry* under the transformations $H \to H + c$, $v \to v - c$, simply because in this limit the action is a functional of $H + v$.[2] The same dilation symmetry is possessed by the regularized one-loop

---

[2] We call this symmetry dilation symmetry since v is the only dimensionful parameter of the theory. This symmetry is violated by the gauge-fixing term Eq. (8.21) in its non-abelian version; this however does not contribute to physical amplitudes, and in particular to the multiphotons amplitudes relevant here.

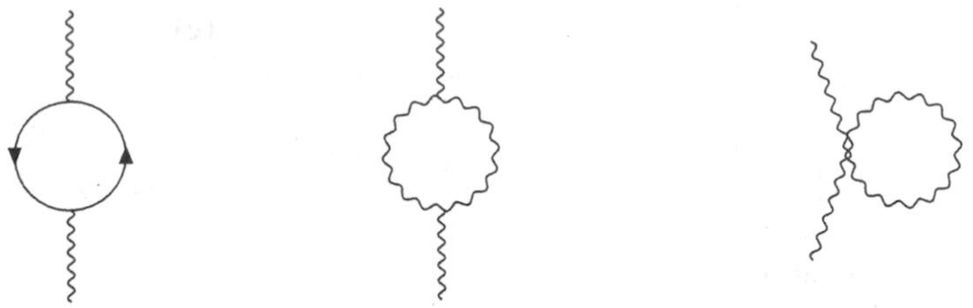

Fig. 12.2 One-loop vacuum-polarization diagrams

effective action. For our purposes, the adopted regularization must preserve electromagnetic gauge invariance, since otherwise loop corrections to the effective action would be functionals of A^μ rather than $F_{\mu\nu}$ alone. As a consequence, the regularized effective action contains a term

$$\begin{aligned} S^{\gamma\gamma}_{\text{eff}}(M) &= -\frac{1}{4}\int d^4x\, \gamma(H(x)+v, M) F^{\mu\nu}(x)F_{\mu\nu}(x) \\ &= -\frac{1}{4}\int d^4x \left[\gamma(v, M) + \frac{\partial\gamma(v, M)}{\partial v} H(x) + \cdots\right] F^{\mu\nu}(x)F_{\mu\nu}(x), \end{aligned} \tag{12.5}$$

where M is the cut-off mass. At one loop, $\gamma(v, M)$ is given by the low-momentum value of the sum of the regularized diagrams in Fig. 12.2, which represent the one-loop corrections to the $\gamma\gamma$ effective vertex, often referred to as the *vacuum polarization*. By gauge invariance, the vacuum polarization has the tensor structure

$$\begin{aligned} \Pi^{\mu\nu}(k, M) &= (k^2 g^{\mu\nu} - k^\mu k^\nu)\Pi(k^2, M) \\ &= (k^2 g^{\mu\nu} - k^\mu k^\nu)\left[\gamma(v, M) + O\left(\frac{k}{v}, \frac{k}{M}\right)\right]. \end{aligned} \tag{12.6}$$

In the first diagram, all charged standard model fermions circulate in the loop, while in the second and third diagrams the internal lines are charged bosons. Correspondingly, we may write

$$\gamma(v, M) = \gamma_f(v, M) + \gamma_W(v, M). \tag{12.7}$$

The coefficient A in Eq. (12.4) is now immediately read off Eq. (12.5):

$$A = \frac{\partial\gamma(v, M)}{\partial v}, \tag{12.8}$$

and it is easily obtained once $\gamma(v, M)$ has been computed. In the renormalized theory, Eq. (12.5) must be supplemented by suitable counter-terms, because $\gamma(v, M)$ diverges in the ultraviolet limit. The renormalization prescriptions require that radiative corrections to the coefficient of $\int dx\, F^{\mu\nu}(x)F_{\mu\nu}(x)$ vanish. Still, $S^{H\gamma\gamma}_{\text{eff}}$ is not affected by counter-terms, because it has mass dimension -1 and hence, according to the dimensional analysis presented in Sect. 4.5, the relevant one-loop diagrams are finite.

This shows that the dilation symmetry is broken beyond the semi classical approximation by the ultraviolet divergences, more precisely by the renormalization counter-terms. In general, counter-terms depend on the internal particle masses, and hence on v, but not on $H(x)$.

Therefore our program consists in computing the unrenormalized one-loop corrections to the electromagnetic vacuum polarization, regularized by the gauge invariant Pauli-Villars method introduced in Sect. 4.5, and identifying the coefficient A in Eq. (12.4) with the v-derivative of the resulting low-momentum coefficient, which is expected to be cut-off independent.

We will employ the dispersion relation approach outlined in Sect. 4.5: we shall first compute the imaginary part of the contribution of each charged particle pair to the $\gamma\gamma$ effective amplitude, which is related to the probability per unit time of the production of the pair by a time-dependent electromagnetic field. This quantity is obviously gauge invariant. Then we shall write the cut-off dispersion relation Eq. (4.62) for the vacuum polarization, omitting the counter-term $C(M^2)$. This is also gauge invariant, because its imaginary part is. Finally, we shall compute the partial derivative of the coefficient $\gamma(v, M)$ with respect to v.

12.2.1 The Matter Fermion Contribution

We begin by computing the contribution from matter fermions f in the loop. To this purpose, we must compute the probability per unit time of the production of a generic $f\bar{f}$ pair by a time-dependent electromagnetic field. The invariant amplitude is given by

$$\mathcal{M}_f = q_f \epsilon_\mu \bar{u}(\boldsymbol{p})\gamma^\mu v(\boldsymbol{k}-\boldsymbol{p}), \tag{12.9}$$

where q_f is the fermion charge and ϵ is the electromagnetic polarization vector. The imaginary part of the fermion contribution to the $\gamma\gamma$ effective amplitude is immediately found using Eqs. (4.49) and (4.21):

$$\begin{aligned} 2\text{Im}\,\Pi_f^{\mu\nu} &= q_f^2 \int d\phi_2 \text{Tr}\left[\gamma^\mu(\not{k}-\not{p}+m_f)\gamma^\nu(\not{p}-m_f)\right] \\ &= 4q_f^2 \int d\phi_2 \left[p^\mu(k-p)^\nu + (k-p)^\mu p^\nu - g^{\mu\nu} p\cdot k\right] \\ &= F(k^2)(k^2 g^{\mu\nu} - k^\mu k^\nu), \end{aligned} \tag{12.10}$$

where $m_f = \frac{h_f v}{\sqrt{2}}$ is the fermion mass. The equality in the last line follows from $k_\mu \text{Im}\,\Pi_f^{\mu\nu} = 0$. This is a consequence of gauge invariance, which can be checked explicitly on the result in the second line of Eq. (12.10). The function $F(k^2)$ can be obtained by taking the trace:

$$\begin{aligned} 3k^2F(k^2) &= 4q_f^2 \int d\phi_2 \left(-2p\cdot k - 2p^2\right) \\ &= -\frac{q_f^2}{\pi}\left(m_f^2 + \frac{k^2}{2}\right)\sqrt{1-\frac{4m_f^2}{k^2}}, \end{aligned} \tag{12.11}$$

where we have used $(p-k)^2 = m_f^2$ and the volume of the two-body phase space obtained from Eqs. (4.25, 4.26). Hence

$$\text{Im}\, \Pi_f^{\mu\nu} = -\frac{q_f^2}{12\pi}(k^2 g^{\mu\nu} - k^\mu k^\nu)\left(1 + \frac{2m_f^2}{k^2}\right)\sqrt{1 - \frac{4m_f^2}{k^2}}. \tag{12.12}$$

We now consider the cut-off dispersion relation for the fermion contribution to the $\gamma\gamma$ amplitude. In analogy with Eq. (4.62), and omitting the counter-term $C(M^2)$, we get

$$\begin{aligned} \Pi_f^{\mu\nu}(M) = -\frac{q_f^2}{12\pi^2}(k^2 g^{\mu\nu} - k^\mu k^\nu) & \left[\int_{4m_f^2}^{4M^2} \frac{d\sigma}{\sigma - k^2 - i\epsilon}\left(1+\frac{2m_f^2}{\sigma}\right)\sqrt{1-\frac{4m_f^2}{\sigma}}\right. \\ & \left. + \int_{4M^2}^{\infty} \frac{d\sigma}{\sigma - k^2 - i\epsilon}\left[\left(1+\frac{2m_f^2}{\sigma}\right)\sqrt{1-\frac{4m_f^2}{\sigma}} - \left(1+\frac{2M^2}{\sigma}\right)\sqrt{1-\frac{4M^2}{\sigma}}\right]\right]. \end{aligned} \tag{12.13}$$

For $M^2 \gg m_f^2, k^2$ the second integral in the square bracket is given by

$$\begin{aligned} & \int_{4M^2}^{\infty} \frac{d\sigma}{\sigma - k^2 - i\epsilon}\left[\left(1+\frac{2m_f^2}{\sigma}\right)\sqrt{1-\frac{4m_f^2}{\sigma}} - \left(1+\frac{2M^2}{\sigma}\right)\sqrt{1-\frac{4M^2}{\sigma}}\right] \\ & \simeq \int_{4M^2}^{\infty} \frac{d\sigma}{\sigma}\left[1 - \left(1+\frac{2M^2}{\sigma}\right)\sqrt{1-\frac{4M^2}{\sigma}}\right] \\ & \simeq \int_0^1 \frac{dx}{x}\left[1 - \left(1+\frac{x}{2}\right)\sqrt{1-x}\right] \end{aligned} \tag{12.14}$$

which is independent of v and M, and can therefore be ignored. Hence

$$\Pi_f^{\mu\nu}(M) = -\frac{q_f^2}{12\pi^2}(k^2 g^{\mu\nu} - k^\mu k^\nu) \int_{\frac{m_f^2}{M^2}}^{1} \frac{dx}{x} \frac{1}{1 - \frac{xk^2}{4m_f^2}} \sqrt{1-x}\left(1 + \frac{x}{2}\right)$$

$$\simeq \frac{q_f^2}{6\pi^2}(k^\mu k^\nu - g^{\mu\nu} k^2) \log \frac{M}{m_f} \qquad (12.15)$$

up to terms that are either negligible as $M \to \infty$, or suppressed by powers of $\frac{k^2}{m_f^2}$. Comparing with Eq. (12.6) we finally obtain

$$\frac{\partial \gamma_f(v, M)}{\partial v} = \frac{q_f^2}{6\pi^2 v}. \qquad (12.16)$$

In the limit of small Higgs mass, $\frac{m_H^2}{4m_f^2} \ll 1$, the coefficient of the effective action term in Eq. (12.8), which has the dimension of m^{-1} and is proportional to the Yukawa coupling $\frac{m_f}{v}$, is necessarily proportional to v^{-1}. Indeed, in the limit, the first diagram in Fig. 12.1 only depends on the fermion mass which therefore fixes the order of magnitude of the loop momentum. If, on the contrary, $\frac{m_H^2}{4m_f^2} \gg 1$ the order of magnitude of the momentum flowing in the loop is given by m_H. However, if m_f vanishes, the region of small loop momenta becomes important: the loop integration has a logarithmic infrared divergence, cut-off by the fermion mass. Therefore the coefficient of the effective action term in this limit is of order

$$\frac{q_f^2}{v} \frac{m_f}{m_H} \log^2 \frac{m_H^2}{4m_f^2} \ll \frac{q_f^2}{v}. \qquad (12.17)$$

Now all the charged matter fermions in the electroweak model, except the top quark in its three color states, satisfy the inequality $\frac{m_H^2}{4m_f^2} \gg 1$ and hence their contribution is negligible. Since the sum of the top squared charges is equal to $\frac{4}{3}e^2$, where e is the proton charge, the top contribution to the effective action coefficient is approximately $\frac{2e^2}{9\pi^2 v}$. Notice in particular that for $m_H = 125$ GeV the order of magnitude of the corrections to the coefficient due to the finite value of the Higgs mass should be $\frac{m_H^2}{4m_f^2} \sim 0.13$.

12.2.2 The W Contribution

The W contribution to the $H\gamma\gamma$ effective vertex in Eq. (12.8) corresponds to the second and third diagrams in Fig. 12.1. Also in this case, one can exploit the dilation symmetry in the small-m_H approximation by computing the contribution to the electromagnetic vacuum polarization of charged scalar and vector particles, that is, the second and third diagrams in Fig. 12.2.

In Sect. 8.4 we have described the Higgs mechanism in the standard model. We have seen that the unitary gauge-fixing choice is the simplest one for tree-level calculations. We have also shown that the unitary gauge has an important drawback, namely the loss of manifest renormalizability, originated by the bad ultraviolet behavior of vector propagators. In the dispersion relation approach to radiative corrections, the loss of renormalizability corresponds to strongly ultraviolet divergent imaginary parts, which in turn implies the need of multiple subtractions; furthermore, the subtraction constants are difficult to determine.

For these reasons, we will perform the calculation with a renormalizable gauge-fixing choice, such as those described in Sect. 10.2. With these choices, the diagrams in Fig. 12.2 receive contributions from the four components of the $W^\pm$ field and the Goldstone boson $G^\pm$. These fields correspond to five degrees of freedom, only three of which are physical. However, as shown in Sect. 10.2, our gauge choice requires the introduction of two more unphysical field multiplets, called the Faddeev-Popov (ghost) fields, which are scalars from the point of view of their Lorentz transformation rules, but fermions form the point of view of statistics: their creation and annihilation operators obey anti-commutation relations. In the case at hand one has two charged ghosts. It turns out that, being quantized with the wrong statistics, ghosts give a negative contribution to the pair production probability by the electromagnetic field, which exactly compensates the positive contribution from the Goldstone bosons and the unphysical, scalar, component of the W. Altogether we remain with three contributing components (degrees of freedom) as expected.[3]

We now turn to the photon coupling to W bosons. In the 't Hooft-Feynman gauge, the $\gamma W^+ W^-$ vertex can be read off the first line of Eq. (10.8), and takes the same form as in Eq. (6.99):

$$V^{\mu\nu\rho}(k,p) = e[g^{\nu\rho}(2p-k)^\mu + g^{\rho\mu}(2k-p)^\nu - g^{\mu\nu}(k+p)^\rho], \qquad (12.18)$$

where we have assigned Lorentz index μ and ingoing momentum k to the photon field, and Lorentz index ν and outgoing momentum p to the W^+, Lorentz index ρ and outgoing momentum $k-p$ to the W^-.

However, this is not the more convenient gauge choice in the present case, because the gauge-fixing term violates electromagnetic gauge invariance. Correspondingly, the mass-shell condition

$$k_\mu V^{\mu\nu\rho}(k,p) = 0 \qquad (12.19)$$

is only obeyed by the first term in the right-hand side of Eq. (12.18) (since $k^\mu(k_\mu - 2p_\mu) = 0$ when $p^2 = (k-p)^2 = m_W^2$), but not by the full vertex. The independence

[3] Notice that in the present framework every field component corresponds to an independent field degree of freedom, thus computing imaginary parts of the vacuum polarization one sums the pair production probabilities of all charged field components.

of the final result of the gauge choice arises in a cumbersome way after the inclusion of Goldstone boson and ghost contributions.

As anticipated at the end of Sect. 10.2, electromagnetic gauge invariance of each separate contribution can be restored by replacing the ordinary derivatives of charged fields in the gauge-fixing term Eq. (10.13) by covariant derivatives, as defined in Eq. (6.15):

$$\partial_\mu W^\mu_\pm \to (\partial_\mu \mp ieA_\mu)W^\mu_\pm = D_\mu W^\mu_\pm. \tag{12.20}$$

This induces similar replacements for the charged ghost and antighost fields $c^\pm$, $\bar{c}^\pm$. To first order in e, these replacements introduce an extra electromagnetic W interaction term

$$ieA_\mu(W^\mu_+\partial_\nu W^\nu_- - W^\mu_-\partial_\nu W^\nu_+), \tag{12.21}$$

which corresponds to an additional term in the vertex in Eq. (12.18)

$$e[g^{\rho\mu}p^\nu - g^{\mu\nu}(k-p)^\rho]. \tag{12.22}$$

The full vertex now reads

$$V_c^{\mu\nu\rho}(k,p) = e[g^{\nu\rho}(2p-k)^\mu + 2(g^{\rho\mu}k^\nu - g^{\mu\nu}k^\rho)], \tag{12.23}$$

where the second term, which we call the *magnetic term*, is gauge invariant even off the mass shell.

A further simplification arises from the fact that the two terms in Eq. (12.23) do not interfere in the calculation of the squared amplitude. This is due to the fact that the first term is symmetric in the W indices, while the second one is antisymmetric.

With the new gauge choice, any electromagnetic coupling to scalar fields (either Goldstone bosons or ghosts) takes the gauge-invariant form

$$V_s^\mu = e(k-2p)^\mu, \tag{12.24}$$

where the positively-charged field is assigned the outgoing momentum p, and the photon is assigned an ingoing momentum k as in the previous cases. Indeed, the vertex function must be proportional to a linear combination of p and k and, up to a constant, $-2p+k$ is the unique combination giving a gauge-invariant vertex. The coefficient e is by definition the charge of the scalar particle, which is fixed by charge conservation.

Notice that the first term in Eq. (12.23) gives, to first order in e, an electromagnetic coupling of each W component, identical (up to a sign) to that of a scalar field. Since with our gauge choice the four W components are mass-degenerate with the Goldstone bosons and with the two Faddeev-Popov ghosts,[4] the sum of scalar and vector contributions to the imaginary part of the vacuum polarization is equal to three

[4] See Sect. 10.2; the modified version of the 't Hooft-Feynman gauge we are adopting here has no impact on scalar masses.

times (4 + 1 − 2) the Goldstone boson contribution, plus the contribution from the magnetic term in the vertex in Eq. (12.23).

12.2.2.1 The Scalar Contribution

We have seen that in the case of a scalar particles the invariant amplitude for the production of a scalar pair is

$$\mathcal{M}_s = e\epsilon_\mu (k - 2p)^\mu, \tag{12.25}$$

where p is the outgoing momentum of the scalar with charge e, and k is the momentum carried by the electromagnetic field.

Using Eqs. (4.49) and (4.21), we find the imaginary part of the scalar contribution to the $\gamma\gamma$ effective amplitude:

$$\begin{aligned} 2\text{Im}\,\Pi_s^{\mu\nu} &= e^2 \int d\phi_2\,(k-2p)^\mu (k-2p)^\nu \\ &= G(k^2)(k^2 g^{\mu\nu} - k^\mu k^\nu), \end{aligned} \tag{12.26}$$

where, as in the case of the spinor contribution, the tensor structure is dictated by gauge invariance, explicitly realized in the first line. We compute $G(k^2)$ by taking the trace:

$$3k^2 G(k^2) = e^2 \int d\phi_2\,(k^2 + 4p^2 - 4p \cdot k) = \frac{e^2}{8\pi}(4m^2 - k^2)\sqrt{1 - \frac{4m^2}{k^2}} \tag{12.27}$$

and therefore

$$\text{Im}\,\Pi_s^{\mu\nu} = \frac{e^2}{48\pi}(k_\mu k_\nu - g_{\mu\nu}k^2)\left(1 - \frac{4m^2}{k^2}\right)^{\frac{3}{2}}. \tag{12.28}$$

Proceeding as in the case of the spinor contribution, we find

$$\begin{aligned} \Pi_s^{\mu\nu}(M) &\simeq \frac{e^2}{48\pi^2}(k_\mu k_\nu - g_{\mu\nu}k^2) \int_{4m^2}^{4M^2} \frac{d\sigma}{\sigma - k^2 - i\epsilon}\left(1 - \frac{4m^2}{\sigma}\right)^{\frac{3}{2}} \\ &\simeq \frac{e^2}{24\pi^2}(k_\mu k_\nu - g_{\mu\nu}k^2)\log\frac{M}{m}. \end{aligned} \tag{12.29}$$

Therefore, using again Eq. (12.6), we have

$$\gamma_s(v, M) = \frac{e^2}{24\pi^2}\log\frac{m}{M}, \tag{12.30}$$

and hence the scalar contribution to the $H \to \gamma\gamma$ coupling is

$$3\frac{d\gamma_s(v, M)}{dv} = \frac{e^2}{8\pi^2 v}\,. \tag{12.31}$$

12.2.2.2 The Magnetic Contribution

It remains to compute the contribution from the second term in the vertex given in Eq. (12.23). For this purpose we consider the production amplitude of a pair $W_\pm$ by a magnetically coupled electromagnetic field. This is

$$\mathcal{M}_m = 2e\epsilon^\mu (g_{\mu\rho}k_\nu - g_{\mu\nu}k_\rho)w_+^\nu w_-^\rho, \tag{12.32}$$

where $w_\pm$ are the polarization vectors of the $W_\pm$ particles. Therefore the imaginary part of the magnetic contribution to the vacuum polarization is given by

$$\begin{aligned} 2\mathrm{Im}\,\Pi_m^{\mu\nu} &= 4e^2(\delta_\rho^\mu k_\sigma - \delta_\sigma^\mu k_\rho)(g^{\nu\rho}k^\sigma - g^{\nu\sigma}k^\rho)\int d\phi_2 \\ &= 8e^2(g^{\mu\nu}k^2 - k^\mu k^\nu)\int d\phi_2 \\ &= \frac{e^2}{\pi}(g^{\mu\nu}k^2 - k^\mu k^\nu)\sqrt{1 - \frac{4m^2}{k^2}}\,. \end{aligned} \tag{12.33}$$

Thus,

$$\begin{aligned} \Pi_m^{\mu\nu}(M) &\simeq \frac{e^2}{2\pi^2}(g^{\mu\nu}k^2 - k^\mu k^\nu)\int_{4m^2}^{4M^2} \frac{d\sigma}{\sigma - k^2 - i\epsilon}\sqrt{1 - \frac{4m^2}{\sigma}} \\ &\simeq \frac{e^2}{\pi^2}(g^{\mu\nu}k^2 - k^\mu k^\nu)\log\frac{M}{m} \end{aligned} \tag{12.34}$$

and

$$\frac{d\gamma_m(v, M)}{dv} = -\frac{e^2}{\pi^2 v}. \tag{12.35}$$

Overall, the contribution to $\gamma(v, M)$ from the gauge sector

$$\gamma_{\mathrm{W}}(v, M) = 3\gamma_s(v, M) + \gamma_m(v, M) \tag{12.36}$$

gives

$$\frac{d\gamma_{\mathrm{W}}(v, M)}{dv} = -\frac{7e^2}{8\pi^2 v}. \tag{12.37}$$

Our final result for the effective $H\gamma\gamma$ coupling in the limit of vanishing Higgs mass is

$$\frac{d\gamma(v, M)}{dv} = -\frac{17e^2}{24\pi^2 v}. \tag{12.38}$$

Corrections to this result originate from the non-vanishing value of the Higgs mass, and can be expressed as a function of the ratio $\tau = \frac{m_H^2}{4m^2}$, where m is either the top quark mass or the W mass. Notice that $\tau_{\text{top}} \sim 0.13$, while one has $\tau_{\text{W}} \sim 0.6$ for $m_H = 125$ GeV, the measured value. It is clear then that the main correction comes from the W contribution, and can only be estimated through a full calculation of the diagrams in Fig. 12.1. For the W contribution one has a correction factor of

$$1 + \frac{22\tau}{105} + \frac{76\tau^2}{735} + O(\tau^3) \simeq 1.163. \tag{12.39}$$

The top contribution correction is much smaller since both τ_{top} and the top contribution to the effective vertex are smaller.

We conclude that the calculation of the $H\gamma\gamma$ effective coupling in the dilation-invariant limit provides a good approximation to the exact value. The calculation of the corresponding decay rate is left to the reader.

12.3 Anomalies

Loop corrections induce a variety of new couplings, beyond those present in the Lagrangian. Some of these are of great physical interest, such as the $H\gamma\gamma$ effective vertex discussed in the previous section. Others, instead, are potentially dangerous. Consider for example the effective vertices that couple either the neutral Goldstone field G, or the scalar component of the Z field, $\partial^\mu Z_\mu$, to two photons. If these couplings were indeed present in the effective action, both G and ∂Z would appear as intermediate states in the $\gamma\gamma \to \gamma\gamma$ scattering amplitude (often referred to as light-light scattering). In Sect. 8.3 we have discussed the analogous situation of the scalar Compton scattering in the abelian Higgs model, where the intermediate states of the Goldstone particle and of the scalar component of the vector particle are present at tree level but cancel each other. If this cancellation ceased to work beyond the tree approximation, the model would present an insuperable physical problem at the level of loop corrections which is usually called an anomaly. Indeed its unitarity would be spoiled, since, as explained in Chap. 8, non-physical states must be excluded either as asymptotic states, or as intermediate states.

In the electroweak model, G and ∂Z do not contribute as intermediate states to light-light scattering in the tree approximation. They might contribute at the level of quantum corrections and, in the absence of cancellations, they would produce an anomaly.

In this Section, we will show that such anomalous contributions may indeed be present, and we shall see under which conditions they cancel in physical amplitudes. To this purpose, we must consider both the $G\gamma\gamma$ and $(\partial Z)\gamma\gamma$ effective amplitudes for a generic choice of the parameters. However, since our purpose is just to show how an anomaly might appear, here we shall choose the parameters which better simplify the calculations. In much the same way as in the $H\gamma\gamma$ case the better choice of the parameters is the one which minimizes the external momenta. In this limit, the Z boson mass is negligible with respect to charged fermion masses. We shall use the same gauge choice as in the case of the $H\gamma\gamma$ calculation.

In the low momentum region, the $G\gamma\gamma$ coupling corresponds to the effective action term

$$\mathcal{A}_{\text{eff}}^{G\gamma\gamma} = \frac{1}{4} B \int d^4x\, G(x) \epsilon_{\mu\nu\rho\sigma} F^{\mu\nu}(x) F^{\rho\sigma}(x), \tag{12.40}$$

since the scalar field G has the opposite parity of H. If this effective coupling appeared after loop corrections, it would induce transitions between two-photon states with orthogonal polarizations in their center-of-mass system.

The effective coupling of the scalar component of Z to the same two gamma state would contain at least one more derivative, e.g. the term $\int dx\, \partial Z(x) \epsilon_{\mu\nu\rho\sigma} F^{\mu\nu}(x) F^{\rho\sigma}(x)$. For this reason in the $m_Z \to 0$ limit the contribution of the scalar component of Z intermediate state vanishes faster than that of G, and hence, if $B \neq 0$ in the same limit, a compensation of the two contributions is excluded.

We now turn to the computation of the coefficient B in Eq. (12.40), that is the coefficient of the anomaly. At the level of one-loop diagrams, the effective $G\gamma\gamma$ vertex receives contributions from triangle diagrams, where one of the charged matter fermions in the theory circulates in the loop. The couplings of G to spinor fields can be read off Eq. (10.18):

$$-iG \sum_f \left(m_{\mathrm{D}}^f \bar{d}^f \gamma_5 d^f - m_{\mathrm{U}}^f \bar{u}^f \gamma_5 u^f + m_{\mathrm{E}}^f \bar{e}^f \gamma_5 e^f \right). \tag{12.41}$$

Hence, the triangular loop diagram has a γ_5 vertex corresponding to the G coupling, and two electromagnetic vertices. Assigning the two photons outgoing momenta q and q' and polarization indices μ and ν respectively, one has, for the generic spinor field s, two diagrams related to one another by the exchange of the final photons. The amplitude of the first diagram is

$$\begin{aligned} &A_1^{\mu\nu}(q, q') \\ &= iK_s \int \frac{d^4k}{(2\pi)^4} \frac{\mathrm{Tr}\,[\gamma_5(\not{k} - \not{q} + m_s)\gamma^\mu(\not{k} + m_s)\gamma^\nu(\not{k} + \not{q}' + m_s)]}{((k-q)^2 - m_s^2 + i\eta)(k^2 - m_s^2 + i\eta)((k+q')^2 - m_s^2 + i\eta)} \\ &= -\frac{4m_s K_s \epsilon^{\mu\nu\rho\sigma}}{(2\pi)^4} q_\rho q'_\sigma \int \frac{d^4k}{((k-q)^2 - m_s^2 + i\eta)(k^2 - m_s^2 + i\eta)((k+q')^2 - m_s^2 + i\eta)}, \end{aligned} \tag{12.42}$$

where $K_s = \pm i m_s q_s^2$, q_s is the spinor field charge, and the negative sign applies to e^f and d^f, while the positive sign applies to u^f. The loop integral is easily computed using the identities

$$\frac{1}{\prod_{i=1}^{n}(A_i + i\eta)} = (-i)^n \int_0^\infty dt\, t^{n-1} \prod_{i=1}^{n} dx_i \,\delta(1 - \sum_{i=1}^{n} x_i)\, e^{it(\sum_{i=1}^{n} x_i A_i + i\eta)}$$

$$\int \frac{d^4k}{(2\pi)^4} e^{itk^2} = \frac{i}{(4\pi i t)^2}$$

$$\int_0^\infty dt\, t^{z-1} e^{it(Y+i\eta)} = i^z \Gamma(z)(Y + i\eta)^{-z}. \tag{12.43}$$

The calculation is further simplified if the mass-shell conditions $q^2 = q'^2 = 0$ for the photons are enforced. We find

$$\begin{aligned} A_s^{\mu\nu}(q, q') &= \left[A_1^{\mu\nu}(q, q') + A_1^{\nu\mu}(q', q)\right]_{q^2 = q'^2 = 0} \\ &= \frac{i 8 m_s K_s}{(4\pi)^2} \epsilon^{\mu\nu\rho\sigma} q_\rho q'_\sigma \int_0^1 dx \int_0^{1-x} dy \frac{1}{xy(q+q')^2 - m_s^2 + i\eta}. \end{aligned} \tag{12.44}$$

In the limit of vanishing q and q' this gives

$$A_s^{\mu\nu}(q, q') \underset{q, q' \to 0}{\to} -\frac{i K_s}{(2\pi)^2 m_s} \epsilon^{\mu\nu\rho\sigma} q_\rho q'_\sigma = \pm \frac{q_s^2}{(2\pi)^2} \epsilon^{\mu\nu\rho\sigma} q_\rho q'_\sigma, \tag{12.45}$$

which corresponds to a contribution B_s to the coefficient in Eq. (12.40)

$$B_s = \pm \frac{q_s^2}{(4\pi)^2}. \tag{12.46}$$

Therefore, the total contribution of the family f is due to three up ($q_s = \frac{2}{3}e$) and down quarks ($q_s = -\frac{1}{3}e$) contributing with opposite signs, and to one charged lepton contributing with the same sign as d. Thus

$$B_f = \frac{e^2}{(4\pi)^2}\left[3\left(\frac{4}{9} - \frac{1}{9}\right) - 1\right] = 0. \tag{12.47}$$

We see that spinor fields do not contribute to the anomaly, provided that for each quak family there is a corresponding lepton family.

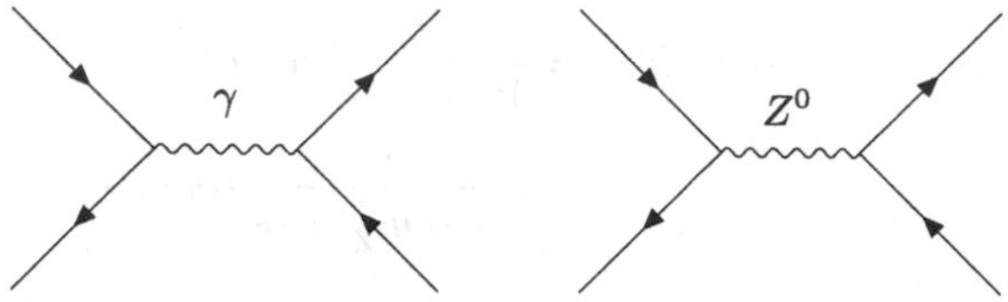

Fig. 12.3 Diagrams for the process $e^+e^- \to \mu^+\mu^-$

Since G also couples to vectors and scalars, one might wonder whether there are other diagrams contributing to B. This possibility is however easily excluded, because scalar and vector couplings cannot produce the antisymmetric tensor structure of Eq. (Eq. (12.40)).

In the general case $(q+q')^2 \neq 0$, the coefficient B_f does not vanish anymore, because the corrections depend on the spinor mass (see Eq. (12.44)). However, in these conditions one must take into account the ∂Z contribution. The calculation is in this case more difficult, but it can be shown that the compensation still takes place.

A similar analysis can be carried on in other cases, such as for example W^+W^- scattering. It can be shown that, under the above-mentioned conditions, the anomaly cancellation still takes place. Furthermore, one may wonder whay happens at higher loops. A general theorem states that, in this particular case of anomalies, cancellation at one loop implies cancellation at all orders.

12.4 The Z^0 Line Shape

A second relevant example of a calculation beyond the semiclassical approximation is the production cross section of a $\mu^+\mu^-$ pair in electron-positron collisions, when the center-of-mass energy is in the neighborhood of the Z^0 mass. In the semiclassical approximation the transition amplitude corresponds to the sum of the two diagrams in Fig. 12.3, where the transition is mediated by a virtual photon or by a Z^0. In the energy region $m_Z - \Delta < \sqrt{s} < m_Z + \Delta$, with Δ of the order of the total Z^0 width, about 2.5 GeV, the contribution of the photon-exchange diagram is about four times smaller than that of the Z^0 diagram. In a precision calculation one should at least take into account the interference between the two contributions. For our present purposes, it will be sufficient to include only the diagram with a Z^0 internal line.

The invariant amplitude is easily obtained by the application of the relevant Feynman rules: the unitary gauge propagator is given in Eq. (10.12), while the vertices are given in Eq. (10.7). Lepton universality implies that muons and electrons are coupled to the Z^0 in the same way. Furthermore, the lepton masses are negligible in the energy regime we are considering. In the center-of-mass reference frame the amplitude is given by

$$\mathcal{M}(e^+e^- \to \mu^+\mu^-) = \frac{g^2}{16\cos^2\theta_W}\bar{v}(-\boldsymbol{p})\gamma_\mu(\gamma_5 - 1 + 4\sin\theta_W)u(\boldsymbol{p})$$
$$\times \frac{g^{\mu\nu}}{s - m_Z^2 + i\epsilon}\bar{u}(\boldsymbol{k})\gamma_\nu(\gamma_5 - 1 + 4\sin\theta_W)v(-\boldsymbol{k}), \quad (12.48)$$

where $\boldsymbol{p}$ is the momentum of the incoming e^-, $\boldsymbol{k}$ the momentum of the outgoing μ^-, and $s = E^2 = 4|\boldsymbol{p}|^2$ the total center-of-mass energy squared. The contribution of the second term in the numerator of the Z^0 propagator vanishes in the limit of massless fermions, because in this limit the corresponding weak neutral currents are conserved. The helicities are not explicitly shown, and will be summed over in the following. The calculation of the square modulus of the invariant amplitude, summed over all fermion helicities, involves the structure

$$X_{\mu\nu}(p, p') = \mathrm{Tr}\left[\not{p}\gamma_\mu(V - A\gamma_5)\not{p}'\gamma_\nu(V - A\gamma_5)\right]$$
$$= 4(V^2 + A^2)(p_\mu p'_\nu + p'_\mu p_\nu - g_{\mu\nu} q \cdot q') - 8iVA\epsilon_{\mu\nu\alpha\beta}p^\alpha p'^\beta \quad (12.49)$$

where, in the present case,

$$V = 1 - 4\sin\theta_W, \qquad A = 1. \quad (12.50)$$

We find

$$\sum_{\text{pol}} |\mathcal{M}(e^+e^- \to \mu^+\mu^-)|^2$$
$$= \left[\frac{g^2}{16\cos^2\theta_W(s - m_Z^2)}\right]^2 X_{\mu\nu}(p, p')X^{\mu\nu}(k', k)$$
$$= 2\left[\frac{g^2}{4\cos^2\theta_W(s - m_Z^2)}\right]^2 [(V^4 + A^4 + 6A^2V^2)(p \cdot k)^2 + (V^2 - A^2)^2(p' \cdot k)^2], \quad (12.51)$$

where $p' = (|\boldsymbol{p}|, -\boldsymbol{p})$, $k' = (|\boldsymbol{k}|, -\boldsymbol{k})$. Denoting by θ the angle between the final muon momentum $\boldsymbol{k}$ and the initial electron momentum $\boldsymbol{p}$ we have

$$\sum_{\text{pol}} |\mathcal{M}(e^+e^- \to \mu^+\mu^-)|^2$$
$$= 2\left[\frac{g^2 s}{16\cos^2\theta_W(s - m_Z^2)}\right]^2 [(V^4 + A^4 + 6A^2V^2)(1 - \cos\theta)^2$$
$$+ (V^2 - A^2)^2(1 + \cos\theta)^2]. \quad (12.52)$$

Then, using Eqs. (4.34) and (4.22), we get the differential cross section

$$
\frac{d\sigma_{e^+e^-\to\mu^+\mu^-}}{d\cos\theta}(s,\cos\theta)
= \frac{s}{64\pi}\left[\frac{g^2}{16\cos^2\theta_W(s-m_Z^2)}\right]^2 [(V^4+A^4+6A^2V^2)(1-\cos\theta)^2
+(V^2-A^2)^2(1+\cos\theta)^2], \tag{12.53}
$$

which gives rise to the familiar phenomenon of forward-backward asymmetry, originated by the presence of an axial coupling. After angular integration we get

$$
\sigma_{e^+e^-\to\mu^+\mu^-}(s) = \frac{s}{12\pi}\left(\frac{g^2}{16\cos^2\theta_W}\right)^2 \frac{(V^2+A^2)^2}{(s-m_Z^2)}. \tag{12.54}
$$

So far, the calculation has been performed at the level of the semi-classical approximation: only tree diagrams have been included. Manifestly, however, the result makes no physical sense since Eq. (12.54) is not integrable over $E=\sqrt{s}$ in any range which includes m_Z, because of the $(E-m_Z)^{-2}$ singularity. We will now see that the paradox is resolved if the loop corrections to the Z^0 propagator are taken into account.

This can be easily seen in the case of a scalar particle. Among the effective vertices introduced in Sect. 12.1 a special role is played by those with two external lines. These effective vertices, denoted by $\gamma_2(p)$, join two propagators with the same momentum p. When p approaches the mass-shell, $p^2 \simeq m^2$, a chain with n insertions of $\gamma_2(p^2)$ enhances the propagator by a factor $\left(-\frac{\gamma_2(p^2)}{m^2-p^2}\right)^n$, which may be large even if $\gamma_2(p^2)$ is perturbatively small. The full chain can be summed over the number of insertions of $\gamma_2(p^2)$, thereby giving a new *full propagator*

$$
\hat{\Delta}(p^2,m^2) = \frac{1}{m^2-p^2-i\epsilon}\sum_{n=0}^{\infty}\left(-\frac{\gamma_2(p^2)}{m^2-p^2-i\epsilon}\right)^n = \frac{1}{m^2-p^2+\gamma_2(p^2)-i\epsilon}. \tag{12.55}
$$

The most apparent consequence of this procedure is that the position of the simple pole of the propagator is shifted by approximately $\gamma_2(m^2)$; this is interpreted as a correction to the squared mass. As already mentioned, in a generic renormalizable theory the two-point effective vertex receives contributions from divergent diagrams which in the case of a scalar field theory are quadratically divergent. Therefore $\gamma_2(p^2)$ is only determined up to a real first order polynomial in p^2, which must be fixed by the renormalization procedure. If $\gamma_2(m^2)$ is real it can be set to zero, together with the first m^2-p^2 Taylor correction, and hence the propagator corrections are scarcely relevant.

However, we have learned in Sect. 4.5 that loop amplitudes have in general complex values. In particular, if the particle is unstable, it can be shown that

$$
\mathrm{Im}\,\gamma_2(m^2) = -m\Gamma, \tag{12.56}
$$

where Γ is the transition rate from the single, unstable, particle state, to all possible multiparticle states. The imaginary part of $\gamma_2(p^2)$ is finite and uniquely identified up to further radiative corrections. Thus, the full propagators of any unstable scalar particle is well approximated, in the region $p^2 \sim m^2$, by

$$\hat{\Delta}(p^2, m^2) \simeq \frac{1}{m^2 - p^2 - im\Gamma}. \tag{12.57}$$

Similar considerations apply to all unstable particles, such as the Higgs particle, or the $W^{\pm}$ and Z^0 gauge bosons. Therefore, we are led to replace $(s - m_Z^2)^2$ with $(s - m_Z^2)^2 + m_Z^2 \Gamma_Z^2$ in Eq. (12.53), where Γ_Z is the total decay probability per unit time of the Z particle at rest:

$$\begin{aligned} \bar{\sigma}_{e^+e^- \to \mu^+\mu^-}(s) &= \frac{s}{12\pi} \left(\frac{g^2}{16 \cos^2 \theta_W} \right)^2 \frac{(V^2 + A^2)^2}{(s - m_Z^2)^2 + m_Z^2 \Gamma_Z^2} \\ &\simeq \frac{1}{48\pi} \left(\frac{g^2}{16 \cos^2 \theta_W} \right)^2 \frac{(V^2 + A^2)^2}{(E - m_Z)^2 + \frac{\Gamma_Z^2}{4}} \end{aligned} \tag{12.58}$$

where the second, approximate expression holds true, for E close to m_Z, if $\Gamma_Z \ll m_Z$, which is in fact the case, since $\Gamma_Z \sim 0.03 m_Z$. This approximation exactly reproduces the Breit-Wigner formula for resonance reactions, whose validity is very general.

We shall compute Γ_Z in the next section. Notice that, if $\Delta \gg \Gamma_Z$,

$$\begin{aligned} \int_{m_Z - \Delta}^{m_Z + \Delta} dE\, \bar{\sigma}_{e^+e^- \to \mu^+\mu^-}(s) &\simeq \int_{-\infty}^{\infty} dE\, \bar{\sigma}_{e^+e^- \to \mu^+\mu^-}(s) \\ &= \frac{1}{24\Gamma_Z} \left(\frac{g^2}{16 \cos^2 \theta_W} \right)^2 (V^2 + A^2)^2. \end{aligned} \tag{12.59}$$

On the contrary, the integral above $m_Z + \Delta$ with $\Delta \gg \Gamma_Z$ gives

$$\int_{m_Z + \Delta}^{\infty} dE\, \bar{\sigma}_{e^+e^- \to \mu^+\mu^-}(s) \simeq \frac{1}{12\pi\Delta} \left(\frac{g^2}{16 \cos^2 \theta_W} \right)^2 (V^2 + A^2)^2. \tag{12.60}$$

12.4.1 The Z^0 Width

The gauge boson Z^0 can only decay into fermion-anti-fermion pairs. In particular, it decays into neutrino and charged lepton pairs of the three known species and into the quarks d, s, b, u and c, each in three color states. The interaction terms responsible for these decays are given in Eq. (10.7). The effects of fermion masses m_f on the decay width are of order $\frac{m_f^2}{m_Z^2}$, which is at most $\sim 3 \times 10^{-3}$ for the b quark. Hence, we shall neglect all fermion masses, which greatly simplifies the calculation.

The invariant amplitude for the Z decay at rest into a generic fermion-antifermion pair $f\bar{f}$ is given by

$$\mathcal{M}_{Z\to f\bar{f}} = \frac{g}{4\cos\theta_W}\epsilon_h^\mu \bar{u}(\boldsymbol{p})\gamma_\mu(V_f - A_f\gamma_5)v(-\boldsymbol{p}). \tag{12.61}$$

Recalling Eq. (4.38), the decay rate of a Z^0 at rest into a $f\bar{f}$ pair is therefore

$$\Gamma_{Z\to f\bar{f}} = \frac{1}{2m_Z}\frac{g^2}{16\cos^2\theta_W}\frac{1}{3}\sum_h \epsilon_h^\mu \epsilon_h^\nu 4(V_f^2 + A_f^2)$$
$$\int (p_\mu p'_\nu + p_\nu p'_\mu - g_{\mu\nu}p\cdot p')\frac{d^3p}{(2\pi m_Z)^2}\delta(m_Z - 2|\boldsymbol{p}|), \tag{12.62}$$

where $p' = (|\boldsymbol{p}|, -\boldsymbol{p})$. In the limit of vanishing fermion mass, $\sum_h \epsilon_h^\mu \epsilon_h^\nu$ can be replaced by $-g^{\mu\nu}$ due to neutral current conservation. Therefore

$$\begin{aligned}\Gamma_{Z\to f\bar{f}} &= \frac{1}{2m_Z}\frac{g^2}{6\cos^2\theta_W}(V_f^2 + A_f^2)\int \frac{d^3p}{(2\pi m_Z)^2}\, p\cdot p'\delta(m_Z - 2|\boldsymbol{p}|)\\ &= \frac{g^2 m_Z}{\cos^2\theta_W}\frac{V_f^2 + A_f^2}{192\pi}.\end{aligned} \tag{12.63}$$

The total decay rate is obtained by summing over the final fermion pairs. We find

$$\Gamma_Z = \frac{\alpha\, m_Z}{48\sin^2\theta_W\cos^2\theta_W}3\left[(V_\nu^2 + A_\nu^2) + (V_e^2 + A_e^2) + 3(V_d^2 + A_d^2) + 2(V_u^2 + A_u^2)\right], \tag{12.64}$$

where the couplings V_f, A_f can be read off Eq. (10.7): $A_f = 1$ for all fermions, $V_\nu = 1$, $V_e = 4\sin^2\theta_W - 1$, $V_d = \frac{4}{3}\sin^2\theta_W - 1$, $V_u = 1 - \frac{8}{3}\sin^2\theta_W$. Replacing these values in Eq. (12.64) and taking into account the known value of θ_W and of m_Z one obtains

$$\Gamma_Z \simeq 2.3\ \text{GeV}, \tag{12.65}$$

very close to the measured value $\Gamma_Z^{\text{exp}} = 2.4952 \pm 0.0023$ GeV.

A naive remark is here in order. For $\sqrt{s} \simeq m_Z$, one may want to consider the process $e^+e^- \to \mu^+\mu^-$ as the production process of a real Z^0 boson, followed by the decay of the Z^0 into a $\mu^+\mu^-$. In the tree approximation this is an unphysical process, because the cross section for $e^+e^- \to Z^0$ is non-zero only for $\sqrt{s} = m_Z$. Indeed, using Eqs. (4.34, 4.35), we find

$$\sigma_{e^+e^-\to Z} = \frac{1}{2s}\frac{1}{4}\sum_{\text{pol}_i, h_f} |\mathcal{M}_{Z\to e^+e^-}|^2 2\pi \int \frac{d^3P}{2E}\,\delta(\boldsymbol{P})\delta(\sqrt{s}-m_Z)$$
$$= \frac{\pi}{m_Z}\frac{g^2}{32\cos^2\theta_W}(V_e^2 + A_e^2)\delta(\sqrt{s}-m_Z). \qquad (12.66)$$

Still, the integral of this cross section over the initial energy E, times the probability of the decay $Z^0 \to \mu^+\mu^-$, which is given by $\frac{\Gamma_{Z\to\mu^+\mu^-}}{\Gamma_Z}$, gives

$$\frac{\Gamma_{Z\to\mu^+\mu^-}}{\Gamma_Z}\int dE\,\sigma_{e\bar{e}\to Z} = \frac{1}{\Gamma_Z}\frac{g^2 m_Z}{\cos^2\theta_W}\frac{V_e^2 + A_e^2}{192\pi}\frac{\pi}{m_Z}\frac{g^2}{32\cos^2\theta_W}(V_e^2 + A_e^2)$$
$$= \frac{1}{24\Gamma_Z}\left(\frac{g^2}{16\cos^2\theta_W}\right)^2 (V_e^2 + A_e^2)^2, \qquad (12.67)$$

which is the correct result, Eq. (12.59).

It is worth mentioning that the identification of the reaction as the production of the Z^0 particle followed by its decay into $\mu^+\mu^-$ decay corresponds to the compound nucleus model which gives the bases for the reaction analyses in nuclear physics. In this model the process is thought to proceed in two steps, the first one being the formations of an unstable compound nucleus state, followed by a decay, typically via tunnel emission, into the final state.

Chapter 13
Neutrino Masses and Mixing

In the original formulation of the standard model, presented in Chaps. 7, 8 and 9, neutrinos are massless particles. This feature is well motivated by direct experimental upper bounds on neutrino masses:

$$m_{\nu_e} \leq 3 \text{ eV}; \quad m_{\nu_\mu} \leq 0.19 \text{ MeV}; \quad m_{\nu_\tau} \leq 18.2 \text{ MeV}, \tag{13.1}$$

and even if observations indicate that neutrino masses are in fact non-zero, the approximation $m_\nu \ll m_f$, where f is any fermion in the standard model spectrum, is extremely good for most applications. In view of experimental results, however, it is interesting to study the possible ways neutrino mass terms can be consistently introduced.

The absence of neutrino mass terms in the standard model is related to the absence of right-handed components for the neutrino fields; these would be assigned (as any other right-handed fermion) to the singlet representation of $SU(2)$, and would have zero charge and hypercharge. Therefore, the corresponding particles do not participate in electroweak gauge interactions, and can be simply omitted in the context of the standard model. One may nevertheless assume that right-handed neutrinos do exist. This assumption brings us outside the standard model, and has far-reaching consequences. Assuming for the moment the existence of only one lepton generation, we introduce a right-handed neutrino through the term

$$\bar{\nu}_R \, i \not{D} \nu_R \equiv \bar{\nu}_R \, i \not{\partial} \nu_R. \tag{13.2}$$

In the presence of a right-handed neutrino field, a Dirac mass term is generated through the Higgs mechanism by a Yukawa coupling similar to that of up-type quarks:

C. M. Becchi and G. Ridolfi, *An Introduction to Relativistic Processes and the Standard Model of Electroweak Interactions*, UNITEXT for Physics,
DOI: 10.1007/978-3-319-06130-6_13,

$$-h_{\mathrm{N}}\left[\bar{\ell}_L\left(\phi+\frac{v}{\sqrt{2}}\right)_c \nu_R+\bar{\nu}_R\left(\phi+\frac{v}{\sqrt{2}}\right)_c^{\dagger} \ell_L\right], \tag{13.3}$$

that contains a term

$$-m\left(\bar{\nu}_L \nu_R+\bar{\nu}_R \nu_L\right); \quad m=\frac{h_{\mathrm{N}} v}{\sqrt{2}} \tag{13.4}$$

in full analogy with the case of quarks. As observed in Sect. 9.3, the Yukawa coupling in Eq. (13.3) cannot be generated by radiative corrections, because it breaks explicitly the global accidental symmetry

$$\nu_R \rightarrow e^{i\phi}\, \nu_R \tag{13.5}$$

of the kinetic term Eq. (13.2), which is the only other place in the Lagrangian density where ν_R appears.

If Eq. (13.4) were the only possible neutrino mass term, then the constant h_{N} should be smaller than the corresponding constants for charged leptons by several order of magnitudes, in order to obey the severe bounds Eq. (13.1); for example,

$$\frac{h_{\mathrm{N}}}{h_e}=\frac{m}{m_e} \sim 10^{-6}. \tag{13.6}$$

In general, however, this is not the case. Because of its transformation properties with respect to gauge transformations, right-handed neutrinos also admit a Majorana mass term of the kind presented in the second line of Eq. (5.71). Here we write this term in the Dirac spinor representation in which the Weyl spinor ξ_R is denoted, as above, by ν_R and, in agreement with Eq. (5.62), $\epsilon\xi_R^*$ is denoted by

$$\nu_R^c \equiv i\gamma_2\gamma_0\bar{\nu}_R^T, \tag{13.7}$$

the charge- conjugated spinor. It is easy to check that the new form of the Majorana mass term is

$$-\frac{1}{2} M\left(\bar{\nu}_R^c \nu_R+\bar{\nu}_R \nu_R^c\right). \tag{13.8}$$

As already observed in Sect. 5.2, this interesting feature is not shared by any other fermion field in the standard model, because of the limitations imposed by gauge invariance. In particular, it is not possible to build a Majorana mass term for left-handed neutrinos by means of renormalizable terms in the Lagrangian density. The Majorana mass parameter M, contrary to the Dirac mass m, can assume arbitrarily large values, since no extra symmetry is recovered in the limit $M=0$. Furthermore, Majorana mass terms violate lepton number conservation; thus, we must assume that M is large enough, in order that lepton number violation effects, typically suppressed by inverse powers of M, are compatible with observations. It is natural to assume that M is of the order of the energy scale characteristic of the unknown phenomena (e.g. the effects of grand unification) experienced by right-handed neutrinos.

The most general neutrino mass term can therefore be written in the form

$$\mathcal{L}_{\nu\ \mathrm{mass}} = -\frac{1}{2}\left(\bar{\nu}_L^c\ \bar{\nu}_R\right)\begin{pmatrix} 0 & m \\ m & M \end{pmatrix}\begin{pmatrix} \nu_L \\ \nu_R^c \end{pmatrix} + \mathrm{h.c.}, \tag{13.9}$$

where we have used $\bar{\nu}_L^c\,\nu_R^c = \bar{\nu}_R\,\nu_L$. We have seen in Sect. 5.2 that the mass matrix that appears in Eq. (13.9) can be written in the diagonal Majorana form

$$\mathcal{L}_{\nu\ \mathrm{mass}} = -\frac{1}{2}\left(\bar{\nu}_1^c\ \bar{\nu}_2\right)\begin{pmatrix} m_1 & 0 \\ 0 & m_2 \end{pmatrix}\begin{pmatrix} \nu_1 \\ \nu_2^c \end{pmatrix} + \mathrm{h.c.}, \tag{13.10}$$

with

$$\nu_1 = i(\cos\theta\,\nu_L - \sin\theta\,\nu_R^c); \quad \nu_2 = \cos\theta\,\nu_R + \sin\theta\,\nu_L^c; \quad \tan 2\theta = \frac{2m}{M} \tag{13.11}$$

and

$$m_1 = \frac{1}{2}\left(\sqrt{M^2+4m^2} - M\right); \quad m_2 = \frac{1}{2}\left(\sqrt{M^2+4m^2} + M\right). \tag{13.12}$$

If, as mentioned above, $m \ll M$, one has

$$\theta \simeq \frac{m}{M}; \quad m_1 \simeq \frac{m^2}{M}; \quad m_2 \simeq M; \quad \nu_1 \simeq i\nu_L; \quad \nu_2 \simeq \nu_R. \tag{13.13}$$

This mechanism, usually called the *see-saw* mechanism, provides a natural explanation of the observed smallness of neutrino masses: one of the two mass eigenstates in the neutrino sector is extremely heavy, and has no observable effects on physics at the weak scale, while the other one has a mass which is suppressed with respect to typical fermion masses m by a factor m/M. In this way, light neutrinos arise without the need of assuming unnaturally small values of the Yukawa couplings.

The see-saw mechanism is easily generalized to the general case of n different species of left-handed neutrinos ($n = 3$ according to present knowledge) and an undetermined number k of right-handed neutrinos; in this case, m is a $k \times n$ matrix, and M a $k \times k$ matrix. After diagonalization in the generation space, the phenomenon of flavour mixing takes place in a similar (but not identical) way as in the quark sector. For simplicity, we will consider the case $k = n$, when there are as many right-handed as left-handed neutrinos. In this case, the Yukawa interaction Eq. (13.3) should be modified to account for the generation structure. Specifically, we replace Eq. (9.19) by

$$\mathcal{L}_Y^{\text{lept}} = -\left[\bar{\ell}'_L \left(\phi + \frac{v}{\sqrt{2}}\right) h'_{\text{E}}\, e'_R + \bar{e}'_R \left(\phi + \frac{v}{\sqrt{2}}\right)^{\dagger} h'^{\dagger}_{\text{E}}\, \ell'_L\right]$$
$$-\left[\bar{\ell}'_L \left(\phi + \frac{v}{\sqrt{2}}\right)_c h'_{\text{N}}\, \nu'_R + \bar{\nu}'_R \left(\phi + \frac{v}{\sqrt{2}}\right)_c^{\dagger} h'^{\dagger}_{\text{N}}\, \ell'_L\right], \quad (13.14)$$

where we have introduced an array ${\nu'_R}^i$; $i = 1, \ldots, n$ of right-handed neutrinos, and h'_{N} is a generic complex constant matrix. The Yukawa coupling induces a Dirac mass term for neutrinos which must be added to the Majorana mass terms for right-handed neutrinos to obtain the full neutrino mass term

$$\mathcal{L}_{\text{neut}} = -\frac{v}{2\sqrt{2}}\left[\left(h'_{\text{N}}\right)_{ij} \left(\bar{\nu}'_{Li}\nu'_{Rj} + \bar{\nu}'^{c}_{Rj}\nu'^{c}_{Li}\right) + \left(h'_{\text{N}}\right)^{*}_{ji} \left(\bar{\nu}'_{Ri}\nu'_{Lj} + \bar{\nu}'^{c}_{Lj}\nu'^{c}_{Ri}\right)\right]$$
$$-\frac{1}{2}\left(M_{ij}\bar{\nu}'^{c}_{Ri}\,\nu'_{Rj} + M^{*}_{ji}\bar{\nu}'_{Ri}\,\nu'^{c}_{Rj}\right), \quad (13.15)$$

where M is a matrix in lepton flavour space which, without loss of generality, can be chosen real, diagonal and positive. If the eigenvalues of M are much larger than the absolute values of the elements of the matrix vh'_N the (Majorana) mass terms for light neutrinos are

$$-\frac{1}{2}\left(\mu_{ij}\bar{\nu}^{c}_{Li}\,\nu_{Lj} + \mu^{*}_{ji}\bar{\nu}_{Li}\,\nu^{c}_{Lj}\right), \quad (13.16)$$

where the indices i, j are lepton flavour indices; the light neutrino fields ν_{Li} only approximately coincide with the ν'_{Li} with the same flavour. In analogy with Eq. (13.14), one finds that to a very good approximation

$$\mu \simeq \frac{v^2}{2}\left(h'_{\text{N}}\right)^T M^{-1}\, h'_{\text{N}} = U^{\dagger}\hat{\mu}U^{*}, \quad (13.17)$$

where $\hat{\mu}$ is a diagonal real matrix and the unitary matrix U is such that

$$\mu\mu^{\dagger} = U^{\dagger}\hat{\mu}^2 U. \quad (13.18)$$

The matrix U, called Pontecorvo-Maki-Nagakawa-Sakata (PMNS) matrix, is identified up to three phases associated with the ν'_{Li}'s, which can be freely chosen due to the lepton flavour conservation property of the electroweak Lagrangian. Hence, U depends on six parameters: three angles and three complex phases. It is the leptonic analogous of the CKM matrix, and gives rise to the lepton flavour mixing (and, possibly, CP violation). It is usually parametrized as

$$U = \begin{pmatrix} c_{12}c_{13} & s_{12}c_{13} & s_{13}e^{i\delta} \\ -s_{12}c_{23} - c_{12}c_{23}s_{13}e^{i\delta} & c_{12}c_{23} - s_{12}s_{23}s_{13}e^{i\delta} & s_{23}c_{13}e^{i\delta} \\ s_{12}s_{23} - c_{12}c_{23}s_{13}e^{i\delta} & -c_{12}s_{23} - s_{12}c_{23}s_{13}e^{i\delta} & c_{23}c_{13}e^{i\delta} \end{pmatrix} \begin{pmatrix} 1 & 0 & 0 \\ 0 & e^{i\alpha} & 0 \\ 0 & 0 & e^{i\beta} \end{pmatrix} \quad (13.19)$$

where c_{ij} and s_{ij} stand for $\cos\theta_{ij}$ and $\sin\theta_{ij}$ respectively.

Neutrinos are produced in weak-interaction processes with a definite flavour: for example, β decays of nuclei in the Sun produce electron neutrinos. Denoting flavour eigenstates by Greek indices, and mass eigenstates by Latin indices, we have

$$|\nu_\alpha\rangle = \sum_{i=1}^{n} U^*_{\alpha i}\,|\nu_i\rangle. \tag{13.20}$$

Let us consider a neutrino beam of definite flavour, produced at the origin $L = 0$. Each definite-mass component of the beam propagates at the distance L as

$$|\nu_i(L)\rangle = e^{ip_i L}\,|\nu_i(0)\rangle, \tag{13.21}$$

where

$$p_i = \sqrt{E^2 - \hat{\mu}_i^2} \simeq E - \frac{\hat{\mu}_i^2}{2E}, \tag{13.22}$$

since neutrinos are almost massless. Hence,

$$|\nu_\alpha(L)\rangle \simeq e^{iEL} \sum_{i=1}^{n} U^*_{\alpha i}\,\exp\left(-i\frac{\hat{\mu}_i^2}{2E}L\right)|\nu_i(0)\rangle, \tag{13.23}$$

where E is the energy of the beam, which is assumed monochromatic for the time being.

The probability amplitude of observing the flavour β at a distance L from the source is given by

$$\begin{aligned}\langle\nu_\beta|\nu_\alpha(L)\rangle &\simeq e^{iEL} \sum_{i=1}^{n} U^*_{\alpha i}\,\exp\left(-i\frac{\hat{\mu}_i^2}{2E}L\right)\sum_{j=1}^{n} U_{\beta j}\,\langle\nu_j|\nu_i\rangle \\ &= e^{iEL}\sum_{i=1}^{n}\xi_i^{\alpha\beta}\,e^{-i\epsilon_i L},\end{aligned} \tag{13.24}$$

where we have defined

$$\xi_i^{\alpha\beta} = U^*_{\alpha i}\,U_{\beta i}; \quad \epsilon_i = \frac{\hat{\mu}_i^2}{2E}. \tag{13.25}$$

The corresponding probability is given by

$$P_{\alpha\beta}(L) = \sum_{i=1}^{n}\sum_{j=1}^{n} \xi_i^{\alpha\beta} \xi_j^{*\alpha\beta} e^{i(\epsilon_j-\epsilon_i)L}$$
$$= \delta_{\alpha\beta} - 4\sum_{i=1}^{n}\sum_{j=i+1}^{n} \text{Re}\left(\xi_i^{\alpha\beta}\xi_j^{*\alpha\beta}\right) \sin^2\frac{1}{2}(\epsilon_j-\epsilon_i)L$$
$$-2\sum_{i=1}^{n}\sum_{j=i+1}^{n} \text{Im}\left(\xi_i^{\alpha\beta}\xi_j^{*\alpha\beta}\right) \sin(\epsilon_j-\epsilon_i)L, \qquad (13.26)$$

where we have used the unitarity of U. Observe that $P_{\alpha\beta}$ is unchanged if one replaces $U \to U^*$ and $\alpha \leftrightarrow \beta$:

$$P(\nu_\alpha \to \nu_\beta; U^*) = P(\nu_\beta \to \nu_\alpha; U). \qquad (13.27)$$

On the other hand, CPT invariance implies

$$P(\nu_\beta \to \nu_\alpha; U) = P(\bar{\nu}_\alpha \to \bar{\nu}_\beta; U). \qquad (13.28)$$

Hence,

$$P(\nu_\alpha \to \nu_\beta; U^*) = P(\bar{\nu}_\alpha \to \bar{\nu}_\beta; U), \qquad (13.29)$$

or in other words neutrino oscillation probabilities can only differ from anti-neutrino oscillation probabilities if $U \neq U^*$, which is also a condition for CP violation.

In real situations Eq. (13.26) requires important corrections for three reasons. First, neutrino beams are not monochromatic. This implies that the particles in the beams are not associated with plane waves as in Eq. (13.21). Rather, their states must be described by wave packets whose space extension is approximately $\frac{1}{\Delta E}$, ΔE being the energy resolution of the beam. The components of the wave packets associated with different mass eigenvalues move along the beam with different velocities:

$$|v_i - v_j| \sim \frac{|\epsilon_i - \epsilon_j|}{E}, \qquad (13.30)$$

and hence different components cease overlapping after a distance $D \sim E/(\Delta E|\epsilon_i - \epsilon_j|)$. More precisely, the exponential in the first line of Eq. (13.26) is replaced by

$$e^{i(\epsilon_j-\epsilon_i)L} \to e^{i(\epsilon_j-\epsilon_i)L} e^{-\frac{(\epsilon_j-\epsilon_i)^2(\Delta E)^2 L^2}{8E^2}} \underset{L\to\infty}{\to} \delta_{ij} \qquad (13.31)$$

and therefore neutrino oscillations are damped after a distance which is approximately equal to the oscillation length times $\frac{E}{\Delta E}$. In some instances, e.g. the case of athmospheric neutrinos, $\frac{\Delta E}{E}$ is of order one; in such cases, what is observed is not oscillations, but a continuous monotonic transition between

$$P_{\alpha\beta}(0) = \delta_{\alpha\beta} \quad \text{and} \quad P_{\alpha\beta}(\infty) = \sum_i |\xi_i^{\alpha\beta}|^2. \tag{13.32}$$

In other cases, e.g. in the case of solar neutrinos with energy of order 10 MeV, $\frac{\Delta E}{E}$ is relatively small, but the observer-source distance L is statistically distributed over millions of kilometers. In these situations, oscillations average to zero, and what is in fact measured is $P_{\alpha\beta}(\infty)$ anywhere. Finally, Eq. (13.26) was obtained under the assumption that neutrino propagation takes place in empty space; in principle, there might be sizable corrections due to the interaction of neutrinos with matter.

In the case of solar neutrinos, which are produced in the electron flavour state ($\alpha = e$) and detected in the same flavour state, one finds a flux reduction factor

$$P_{ee}(\infty) = \sum_i |U_{ei}|^4 = 0.58 \pm 0.07 \tag{13.33}$$

in the 1 MeV energy region. A different flux reduction, $P_{ee}(\infty) \simeq 0.3$, is instead measured for solar neutrinos in the 10 MeV energy region. This difference cannot be explained by Eq. (13.32), which is manifestly independent of the neutrino energy.

An elegant explanation of this effect is based on the possibility that oscillations in matter be different from those in empty space. This might look surprising since neutrinos interact very weakly. However, it has been suggested that in certain conditions of neutrino energy and electron density, and for certain values of the relevant neutrino squared mass differences, a resonance mechanism can take place, the so-called the Mikheyev-Smirnov-Wolfenstein (MSW) mechanism, which may modify the first (electron) line of the PMNS matrix setting in particular $U_{e2} = 1$ in matter.[1] In these conditions, electrons would be created in the Sun in a mass eigenstate, and the beam would remain in the same mass eigenstate also emerging from the Sun into the vacuum.[2] In this situation one would find a flux reduction factor equal to the vacuum value of $|U_{e2}|^2$, which might fairly well be close to 1/3. Given the solar electron density and the neutrino energy, the resonance hypothesis favours a squared mass difference $\Delta m_\odot^2 \sim 7 \times 10^{-5}\,\text{eV}^2$.

An important source of experimental information on neutrinos is the study of the multi-GeV atmospheric neutrinos produced by the interactions of primary cosmic rays with the atmosphere. One observes a reduction by a factor about two in the muon neutrino flux when the neutrino azimutal angle varies between zero (particles coming from above) and 180° (particles from below), and hence L varies between few times 10 km and 2×10^4 km. One observes about five damped oscillations corresponding to an L of 13,000 km, the diameter of the earth; for an average energy of about 8 GeV, Eq. (13.26) gives $\Delta m^2 \sim 3 \times 10^{-3}\,(\text{eV})^2$.

---

[1] The basic mechanism is that shown in Sect. 4.4 for a relativistic scalar particle in matter. In the case of solar neutrinos the matter particles are electrons which have different forward scattering amplitudes with the neutrinos of different species due to the presence of charged current interactions.

[2] This is a consequence of the quantum mechanical version of the adiabatic theorem.

The choice of the ordering of the mass eigenstates is, of course, arbitrary. The parametrization of the PMNS mixing matrix given in Eq. (13.19) is motivated by the fact that the solar problem seems to involve two mass eigenstates, which are conventionally identified with the first two eigenstates:

$$\Delta m^2_{\odot} = \Delta m^2_{21}. \tag{13.34}$$

On the other hand, since most of the atmospheric neutrinos (~2/3) are μ neutrinos, it is natural to identify the mass difference Δm^2_A measured in atmospheric neutrino experiments with Δm^2_{31}. Because $\Delta m^2_A \gg \Delta m^2_{\odot}$, we conclude that $\Delta m^2_{31} \simeq \Delta m^2_{32}$.

Further important experiments originate from the anti-neutrino flux generated by the nuclear power stations, which can be measured at distances of few kilometers, and from long baseline experiments based on high-energy artificial (anti)-neutrino beams, which will be crucial in order to detect possible CP violating phases in the PMNS matrix.

The analyses of the anti-neutrino flux generated by the nuclear power stations at distances of the order of one kilometer can put into evidence oscillations in the electron anti-neutrino survival probability $P_{ee}(L)$ corresponding to $\Delta m^2_{31} \simeq \Delta m^2_{32}$. Indeed, using Eqs. (13.26) and (13.19), one has

$$P_{ee}(L) = 1 - \sin^2(2\theta_{13}) \sin^2 \frac{\Delta m^2_{31} L}{4E}, \tag{13.35}$$

and for $E \sim 3\,\text{MeV}$, the tipical average value of antineutrino energy, one has an oscillation length $L \sim 600\,\text{m}$ and a damping length few times larger. Recently, a tiny effect has been detected, which can be interpreted in terms of a small, but non zero, value of $\sin^2\theta_{13} \simeq 0.025$.

Appendix A
Large-Time Evolution of the Free Field

Let us consider a negative-frequency free-field solution of the equation of motion for a real scalar field:

$$F(\boldsymbol{r}, t) = \frac{1}{(2\pi)^{\frac{3}{2}}} \int \frac{d^3p}{\sqrt{2E_p}}\, \phi(\boldsymbol{p}) e^{i(\boldsymbol{p}\cdot\boldsymbol{r}-E_p t)} \equiv \int d^3p\, \psi(\boldsymbol{p})\, e^{i(\boldsymbol{p}\cdot\boldsymbol{r}-E_p t)}, \tag{A.1}$$

where $E_p = \sqrt{\boldsymbol{p}^2 + m^2}$, and $\phi(\boldsymbol{p})$ is a linear combination of wave packets, assumed to be sufficiently regular (typically, of gaussian shape.) For $t \to \pm\infty$ and fixed $\boldsymbol{r}$, the integral vanishes faster than any negative power of t, because the integral over momenta of the phase factor averages to zero. If however the position $\boldsymbol{r}$ is not held fixed, there is in general a region in the space of momenta where the phase factor is stationary; this region is identified by the condition

$$\frac{\partial}{\partial p^i}(\boldsymbol{p}\cdot\boldsymbol{r} - E_p t) = r_i - \frac{p_i}{E_p} t = 0, \tag{A.2}$$

or

$$\boldsymbol{p} = \frac{m\boldsymbol{v}}{\sqrt{1-v^2}}; \qquad \boldsymbol{v} = \frac{\boldsymbol{r}}{t}, \tag{A.3}$$

and gives the dominant contribution to the integral in the large-time limit. Thus, we take $\boldsymbol{r} = \boldsymbol{v}t$ and we consider the limit of Eq. (A.1) for large t at constant $\boldsymbol{v}$:

$$\lim_{t\to\infty} \int d^3p\, \psi(\boldsymbol{p})\, e^{i(\boldsymbol{p}\cdot\boldsymbol{v}-E_p)t}. \tag{A.4}$$

Expanding the phase in powers of $\boldsymbol{q} \equiv \boldsymbol{p} - \frac{m\boldsymbol{v}}{\sqrt{1-v^2}}$ up to second order we get

$$\boldsymbol{p}\cdot\boldsymbol{v} - E_p = -\sqrt{1-v^2}\left[m - \frac{1}{2m}(q^2 - (\boldsymbol{q}\cdot\boldsymbol{v})^2)\right] + O(q^3). \tag{A.5}$$

C. M. Becchi and G. Ridolfi, *An Introduction to Relativistic Processes and the Standard Model of Electroweak Interactions*, UNITEXT for Physics,
DOI: 10.1007/978-3-319-06130-6,

The asymptotic limit of the integral is therefore

$$\int d^3p\,\psi(\boldsymbol{p})\,e^{i(\boldsymbol{p}\cdot\boldsymbol{v}-E_p)t} \underset{t\to\infty}{\sim} e^{-im\sqrt{1-v^2}t}\,\psi\left(\frac{m\boldsymbol{v}}{\sqrt{1-v^2}}\right)\int d^3q\,e^{-i\sqrt{1-v^2}\frac{q^2-(\boldsymbol{q}\cdot\boldsymbol{v})^2}{2m}t}$$

$$= e^{-im\sqrt{1-v^2}t}\,\psi\left(\frac{m\boldsymbol{v}}{\sqrt{1-v^2}}\right)\left(\frac{2\pi m}{it}\right)^{\frac{3}{2}}\frac{1}{(1-v^2)^{\frac{5}{4}}}. \qquad \text{(A.6)}$$

By the same technique, we can obtain the following interesting result:

$$\int d^3p\,e^{-\frac{(\boldsymbol{p}-\boldsymbol{k})^2}{\sigma^2}}\,e^{\pm iE_pt} \underset{t\to\infty}{\sim} e^{\pm imt-\frac{k^2}{2\sigma^2}}\left(\pm\frac{2\pi im}{t}\right)^{\frac{3}{2}}. \qquad \text{(A.7)}$$

Appendix B
The S Matrix

In this Appendix, we give a proof of Eq. (3.18),

$$\mathcal{S}_{qQ}^{nm} = \mathcal{I}_{qQ}^{nm} - 2\pi i\delta(E_q - E_Q)\mathcal{T}_{qQ}^{nm}, \tag{B.1}$$

where

$$\mathcal{S}_{qQ}^{nm} = \langle \Psi_{q,n}^- | \Psi_{Q,m}^+ \rangle \tag{B.2}$$

and

$$|\Psi_{k,n}^\pm\rangle = |\varphi_{k,n}\rangle + \frac{1}{E_k - H_0 \pm i\epsilon} V\, |\Psi_{k,n}^\pm\rangle. \tag{B.3}$$

We will also show that

$$\begin{aligned} -2\pi i\delta\left(E_q - E_Q\right)\mathcal{T}_{qQ}^{nm} &= -2\pi i\delta\left(E_q - E_Q\right)\langle\varphi_{q,n}|V|\Psi_{Q,m}^+\rangle \\ &= -2\pi i\delta\left(E_q - E_Q\right)\langle\Psi_{q,n}^-|V|\varphi_{Q,m}\rangle. \end{aligned} \tag{B.4}$$

We have

$$\begin{aligned} \langle\Psi_{q,n}^-|\Psi_{Q,m}^+\rangle &= \langle\varphi_{q,n} + \frac{1}{E_q - H_0 - i\epsilon}V\Psi_{q,n}^-|\varphi_{Q,m} + \frac{1}{E_Q - H_0 + i\epsilon}V\Psi_{Q,m}^+\rangle \\ &= \mathcal{I}_{qQ}^{nm} + \frac{\langle\varphi_{q,n}|V|\Psi_{Q,m}^+\rangle}{E_Q - E_q + i\epsilon} - \frac{\langle\Psi_{q,n}^-|V|\varphi_{Q,m}\rangle}{E_Q - E_q - i\epsilon} \\ &+ \langle\Psi_{q,n}^-|V\frac{1}{E_q - H_0 + i\epsilon}\frac{1}{E_Q - H_0 + i\epsilon}V|\Psi_{Q,m}^+\rangle, \end{aligned} \tag{B.5}$$

where we have used $H_0\varphi_{q,n} = E_q\varphi_{q,n}$. We now use the identities

$$\frac{1}{x + i\epsilon} = \frac{1}{x - i\epsilon} - 2\pi i\delta(x) \tag{B.6}$$

C. M. Becchi and G. Ridolfi, *An Introduction to Relativistic Processes and the Standard Model of Electroweak Interactions*, UNITEXT for Physics,
DOI: 10.1007/978-3-319-06130-6,

and

$$\frac{1}{E_q - H_0 + i\epsilon}\frac{1}{E_Q - H_0 + i\epsilon} = \frac{1}{E_Q - E_q}\left(\frac{1}{E_q - H_0 + i\epsilon} - \frac{1}{E_Q - H_0 + i\epsilon}\right) \tag{B.7}$$

to obtain

$$\begin{aligned}\langle\Psi^-_{q,n}|\Psi^+_{Q,m}\rangle = \mathcal{I}_{qQ} &- 2\pi i\delta(E_Q - E_q)\langle\varphi_{q,n}|V|\Psi^+_{Q,m}\rangle \\ &+ \frac{1}{E_Q - E_q - i\epsilon}\langle\varphi_{q,n} + \Psi^-_q V\frac{1}{E_q - H_0 + i\epsilon}|V\Psi^+_{Q,m}\rangle \\ &- \frac{1}{E_Q - E_q - i\epsilon}\langle\Psi^-_{q,n}|V\varphi_{Q,m} + V\frac{1}{E_Q - H_0 + i\epsilon}V\Psi^+_{Q,m}\rangle.\end{aligned} \tag{B.8}$$

The last two terms cancel against each other, because they are both proportional to

$$\langle\Psi^-_{q,n}|V|\Psi^+_{Q,m}\rangle. \tag{B.9}$$

Hence

$$\mathcal{S}^{nm}_{qQ} = \mathcal{I}^{nm}_{qQ} - 2\pi i\delta(E_Q - E_q)\langle\varphi_{q,n}|V|\Psi^+_{Q,m}\rangle \tag{B.10}$$

as announced. A similar arguments leads to

$$\mathcal{S}_{qQ} = \mathcal{I}_{qQ} - 2\pi i\delta(E_Q - E_q)\langle\Psi^-_{q,n}|V|\varphi_{Q,m}\rangle. \tag{B.11}$$

Using

$$|\Psi^+_{Q,m}\rangle = |\varphi_{Q,m}\rangle + \frac{1}{E_Q - H_0 + i\epsilon}V\,|\Psi^+_{Q,m}\rangle \tag{B.12}$$

recursively, we get

$$\begin{aligned}\langle\varphi_{q,n}|V|\Psi^+_{Q,m}\rangle = \langle\varphi_{q,n}|V|\varphi_{Q,m}\rangle &+ \langle\varphi_{q,n}|V\frac{1}{E_Q - H_0 + i\epsilon}V|\varphi_{Q,m}\rangle \\ &+ \langle\varphi_{q,n}|V\frac{1}{E_Q - H_0 + i\epsilon}V\frac{1}{E_Q - H_0 + i\epsilon}V|\varphi_{Q,m}\rangle \\ &+ \cdots\end{aligned} \tag{B.13}$$

The second term in this expansion can be written

$$\begin{aligned}&\sum_a\sum_b\langle\varphi_{q,n}|V|\varphi_{p_a,n_a}\rangle\frac{\langle\varphi_{p_a,n_a}|\varphi_{p_b,n_b}\rangle}{E_Q - E_{p_b} + i\epsilon}\langle\varphi_{p_b,n_b}|V|\varphi_{Q,m}\rangle \\ &= \frac{\langle\varphi_{q,n}|V|0\rangle\langle 0|V|\varphi_{Q,m}\rangle}{E_Q + i\epsilon} + \sum_{n_a}\int D^{3n_a}p_a\,\frac{\langle\varphi_{q,n}|V|\varphi_{p_a,n_a}\rangle\langle\varphi_{p_a,n_a}|V|\varphi_{Q,m}\rangle}{E_Q - E_{p_a} + i\epsilon}\end{aligned} \tag{B.14}$$

Appendix C
Spectral Representation for the S Matrix

An alternative approach to the $\mathcal{S}$ matrix, often found in the literature, is based on a spectral representation of the time-evolution operator. Let us consider the quantity

$$\lim_{t\to\infty}\lim_{t'\to-\infty}\langle\varphi_f(t)|e^{-\frac{i}{\hbar}H(t-t')}|\varphi_g(t')\rangle$$
$$=\lim_{t\to\infty}\lim_{t'\to-\infty}\langle\varphi_f|e^{\frac{i}{\hbar}H_0 t}e^{-\frac{i}{\hbar}H(t-t')}e^{-\frac{i}{\hbar}H_0 t'}|\varphi_g\rangle, \tag{C.1}$$

where the free-particle wave packets $|\varphi_{g,f}\rangle$ are defined in Eqs. (3.13) and (3.16). One can construct a spectral representation the time-evolution operator $e^{-\frac{i}{\hbar}H(t-t')}$ in terms of the incoming (outgoing) states in the equivalent forms

$$e^{-\frac{i}{\hbar}H(t-t')}=\sum_m\frac{1}{m!}\int D^{3m}Q\,e^{-\frac{i}{\hbar}E_Q(t-t')}|\Psi^+_{Q,m}\rangle\langle\Psi^+_{Q,m}| \tag{C.2}$$
$$=\sum_m\frac{1}{m!}\sum_n\frac{1}{n!}\int D^{3m}Q\int D^{3n}q\,e^{-\frac{i}{\hbar}E_q t}|\Psi^-_{q,n}\rangle S^{nm}_{qQ}\langle\Psi^+_{Q,m}|e^{\frac{i}{\hbar}E_Q t'}.$$

Using the definition Eq. (3.2) we find, in analogy with Eqs. (3.12) and (3.14),

$$e^{-\frac{i}{\hbar}E_q t}\langle\varphi_f|e^{\frac{i}{\hbar}H_0 t}|\Psi^-_{q,n}\rangle=\int D^{3\ell}p\,f^*(\boldsymbol{p})\langle\varphi_{\boldsymbol{p},\ell}|\Psi^-_{q,n}\rangle e^{\frac{i}{\hbar}(E_p-E_q)t} \tag{C.3}$$
$$=f^*(\boldsymbol{q})\delta_{\ell n}-\frac{i}{\hbar}\int_{+\infty}^{t}d\tau\int D^{3\ell}p\,f^*(\boldsymbol{p})e^{\frac{i}{\hbar}(E_p-E_q)\tau}\langle\varphi_{\boldsymbol{p},\ell}|V|\Psi^-_{q,n}\rangle$$

$$e^{\frac{i}{\hbar}E_Q t'}\langle\Psi^+_{Q,m}|e^{-\frac{i}{\hbar}H_0 t'}|\varphi_g\rangle=\int D^{3\ell}p\,g(\boldsymbol{p})\langle\Psi^+_{Q,m}|\varphi_{\boldsymbol{p},\ell}\rangle e^{\frac{i}{\hbar}(E_Q-E_p)t'} \tag{C.4}$$
$$=g(\boldsymbol{Q})\delta_{m\ell}+\frac{i}{\hbar}\int_{-\infty}^{t'}d\tau\int D^{3\ell}p\,g(\boldsymbol{p})e^{\frac{i}{\hbar}(E_Q-E_p)\tau}\langle\Psi^+_{Q,m}|V|\varphi_{\boldsymbol{p},\ell}\rangle.$$

C. M. Becchi and G. Ridolfi, *An Introduction to Relativistic Processes and the Standard Model of Electroweak Interactions*, UNITEXT for Physics,
DOI: 10.1007/978-3-319-06130-6, © Springer International Publishing Switzerland 2014

As a consequence,

$$\lim_{t\to\infty}\lim_{t'\to-\infty}\langle\varphi_f|e^{\frac{i}{\hbar}H_0t}e^{-\frac{i}{\hbar}H(t-t')}e^{-\frac{i}{\hbar}H_0t'}|\varphi_g\rangle = \int D^{3n_i}QD^{3n_f}q f^*(\boldsymbol{q})g(\boldsymbol{Q})S^{n_in_f}_{\boldsymbol{qQ}}\,. \tag{C.5}$$

This proves the well known result that the operator

$$U(t,t') = e^{\frac{i}{\hbar}H_0t}e^{-\frac{i}{\hbar}H(t-t')}e^{-\frac{i}{\hbar}H_0t'} \tag{C.6}$$

converges weakly to the S matrix when $t\to\infty$ and $t'\to-\infty$.

Appendix D
Transition Amplitudes in the High Resolution Limit

In this Appendix, we will give an expression for the amplitude Eq. (3.23) for a generic scattering process which is more suited for its calculation in the semi-classical approximation. To this purpose, we observe that $\mathcal{S}$-matrix elements can be written in terms of asymptotic particle creation operators, Eq. (3.4):

$$\mathcal{S}_{qQ}^{n2} = \langle\Omega|\prod_{i=1}^{n} A_{\text{out}}(\boldsymbol{q}_i)\prod_{j=1}^{2} A_{\text{in}}^{\dagger}(\boldsymbol{Q}_j)|\Omega\rangle. \tag{D.1}$$

Using the explicit expressions for the wave packets, and taking for simplicity the same momentum resolution δ for both initial-state and final-state wave packets, we find

$$\mathcal{A}_{\boldsymbol{kp}}^{n2} = \left(\frac{1}{\sqrt{\pi}\delta}\right)^{\frac{3}{2}(n+2)} \prod_{i=1}^{n}\int d^3q_i\, e^{-\frac{(\boldsymbol{k}_i-\boldsymbol{q}_i)^2}{2\delta^2}} \prod_{j=1}^{2}\int d^3Q_j\, e^{-\frac{(\boldsymbol{p}_i-\boldsymbol{Q}_j)^2}{2\delta^2}}$$
$$\langle\Omega|\prod_{i=1}^{n} A_{\text{out}}(\boldsymbol{q}_i)\prod_{j=1}^{2} A_{\text{in}}^{\dagger}(\boldsymbol{Q}_j)|\Omega\rangle, \tag{D.2}$$

with

$$\sum_{i=1}^{n}\boldsymbol{k}_i = \sum_{j=1}^{2}\boldsymbol{p}_j, \qquad \sum_{i=1}^{n}E_{\boldsymbol{k}_i} = \sum_{j=1}^{2}E_{\boldsymbol{p}_j}. \tag{D.3}$$

In order to study the the small-δ behaviour of the amplitude, under the assumption that all particle momenta are different, it is convenient to change the integration variables as

$$\boldsymbol{q}_i = \boldsymbol{k}_i + \hat{\boldsymbol{q}}_i\delta; \qquad \boldsymbol{Q}_j = \boldsymbol{p}_j + \hat{\boldsymbol{Q}}_j\delta, \tag{D.4}$$

C. M. Becchi and G. Ridolfi, *An Introduction to Relativistic Processes and the Standard Model of Electroweak Interactions*, UNITEXT for Physics,
DOI: 10.1007/978-3-319-06130-6, © Springer International Publishing Switzerland 2014

so that

$$\mathcal{A}^{n2}_{\boldsymbol{kp}} = \left(\frac{\delta}{\sqrt{\pi}}\right)^{\frac{3}{2}(n+2)} \prod_{i=1}^{n} \int d^3\hat{q}_i\, e^{-\frac{\hat{q}_i^2}{2}} \prod_{j=1}^{2} \int d^3\hat{Q}_j\, e^{-\frac{\hat{Q}_j^2}{2}}$$
$$\langle\Omega| \prod_{i=1}^{n} A_{\text{out}}(\boldsymbol{k}_i + \hat{\boldsymbol{q}}_i\delta) \prod_{j=1}^{2} A^{\dagger}_{\text{in}}(\boldsymbol{p}_j + \hat{\boldsymbol{Q}}_j\delta)|\Omega\rangle. \tag{D.5}$$

The residual dependence of the vacuum expectation value on δ can be made explicit, recalling that it is proportional to an energy-momentum conservation delta function

$$\delta\Big(\sum_{i=1}^{n}(\boldsymbol{k}_i + \hat{\boldsymbol{q}}_i\delta) - \sum_{j=1}^{2}(\boldsymbol{p}_j + \hat{\boldsymbol{Q}}_j\delta)\Big)\delta\Big(\sum_{i=1}^{n} E_{\boldsymbol{k}_i+\hat{\boldsymbol{q}}_i\delta} - \sum_{j=1}^{2} E_{\boldsymbol{p}_j+\hat{\boldsymbol{Q}}_j\delta}\Big)$$
$$= \frac{2}{\delta^4}\delta\Big(\sum_{i=1}^{n}\hat{\boldsymbol{q}}_i - \sum_{j=1}^{2}\hat{\boldsymbol{Q}}_j\Big)\delta\Big(\sum_{i=1}^{n}\boldsymbol{v}_i\cdot\hat{\boldsymbol{q}}_i - \sum_{j=1}^{2}\boldsymbol{V}_j\cdot\hat{\boldsymbol{Q}}_j + O(\delta)\Big), \tag{D.6}$$

where

$$\boldsymbol{v}_i = \frac{\boldsymbol{k}_i}{E_{\boldsymbol{k}_i}}; \qquad \boldsymbol{V}_j = \frac{\boldsymbol{p}_j}{E_{\boldsymbol{p}_j}} \tag{D.7}$$

and we have used Eq. (D.3). Hence

$$\mathcal{A}^{n2}_{\boldsymbol{kp}} \sim \delta^{\frac{3}{2}n-1}, \tag{D.8}$$

which is consistent with our previous result Eq. (3.44).

We now introduce the initial and final creation operators

$$A^{\dagger}_{I} \equiv \frac{1}{\sqrt{2}}\sum_{j=1}^{2} A^{\dagger}_{j,\text{in}}, \qquad A^{\dagger}_{F} \equiv \frac{1}{\sqrt{n}}\sum_{i=1}^{n} A^{\dagger}_{i,\text{out}}, \tag{D.9}$$

where

$$A^{\dagger}_{j,\text{in}} = \int d^3Q\, \psi_{\boldsymbol{p}_j}(\boldsymbol{Q})A^{\dagger}_{\text{in}}(\boldsymbol{Q}); \qquad A^{\dagger}_{i,\text{out}} = \int d^3q\, \psi_{\boldsymbol{k}_i}(\boldsymbol{q})A^{\dagger}_{\text{out}}(\boldsymbol{q}) \tag{D.10}$$

so that

$$\mathcal{A}^{n2}_{\boldsymbol{kp}} = \langle\Omega| \prod_{i=1}^{n} A_{i,\text{out}} \prod_{j=1}^{2} A^{\dagger}_{j,\text{in}}|\Omega\rangle. \tag{D.11}$$

The operators A_I, A_F are normalized so that

$$[A_I, A_I^\dagger] = 1, \qquad [A_F, A_F^\dagger] = 1. \tag{D.12}$$

In the limit of infinite momentum resolution of the wave packets, $\delta \to 0$, the amplitude Eq. (D.11) can be written in terms of the (non normalized) coherent states

$$|I\rangle \equiv e^{\sqrt{2}A_I^\dagger}|\Omega\rangle, \qquad |F\rangle \equiv e^{\sqrt{n}A_F^\dagger}|\Omega\rangle \tag{D.13}$$

as

$$\langle F|I\rangle - 1 = \mathcal{A}_{kp}^{n2}\left[1 + O(\delta^{\frac{3n}{2}-1})\right]. \tag{D.14}$$

Here is the proof. Using the definitions in Eqs. (D.9) and (D.10) we find

$$\begin{aligned}\langle F|I\rangle &= \langle\Omega|e^{\sum_{i=1}^n A_{i,\mathrm{out}}}e^{\sum_{j=1}^2 A_{j,\mathrm{in}}^\dagger}|\Omega\rangle \\ &= \langle\Omega|\prod_{i=1}^n e^{A_{i,\mathrm{out}}}\prod_{j=1}^2 e^{A_{j,\mathrm{in}}^\dagger}|\Omega\rangle \\ &= \sum_{\nu_1,\ldots,\nu_n=0}^\infty \sum_{\mu_1,\mu_2=0}^\infty \frac{1}{\prod_{i=1}^n \nu_i! \prod_{j=1}^2 \mu_j!}\langle\Omega|\prod_{i=1}^n (A_{i,\mathrm{out}})^{\nu_i}\prod_{j=1}^2 (A_{j,\mathrm{in}}^\dagger)^{\mu_j}|\Omega\rangle.\end{aligned} \tag{D.15}$$

Due to vacuum translation invariance, and for sufficiently small δ, the vacuum expectation values (vev's) in the right-hand side of Eq. (D.15) are non-zero only for

$$\sum_{i=1}^n \nu_i \boldsymbol{k}_i = \sum_{j=1}^2 \mu_j \boldsymbol{p}_j, \qquad \sum_{i=1}^n \nu_i E_{\boldsymbol{k}_i} = \sum_{j=1}^2 \mu_j E_{\boldsymbol{p}_j}. \tag{D.16}$$

Now, the constraints Eq. (D.16) are compatible with energy-momentum conservation, Eqs. (D.3), only if the ν_i's and μ_j's are equal to the same integer K. Thus

$$\langle F|I\rangle - 1 = \sum_{K=1}^\infty \frac{1}{(K!)^{n+2}}\langle\Omega|\prod_{i=1}^n (A_{i,\mathrm{out}})^K \prod_{j=1}^2 (A_{j,\mathrm{in}}^\dagger)^K|\Omega\rangle. \tag{D.17}$$

In quantum field theory, vacuum expectations values of products of operators $\langle\Omega|\prod_{j=1}^n O_j|\Omega\rangle$ are recursively decomposed into truncated parts, defined iteratively by the following construction[1]:

---

[1] In the general case, when the O_i's are local operators in a massive theory, the truncated vev's satisfy a cluster property, that is, they vanish exponentially when any space-like distance between the operator points diverges.

$$\begin{aligned}
\langle\Omega|O_i|\Omega\rangle &= \langle\Omega|O_i|\Omega\rangle_T \\
\langle\Omega|O_iO_j|\Omega\rangle &= \langle\Omega|O_iO_j|\Omega\rangle_T + \langle\Omega|O_i|\Omega\rangle_T\langle\Omega|O_j|\Omega\rangle_T \\
\langle\Omega|O_iO_jO_k|\Omega\rangle &= \langle\Omega|O_iO_jO_k|\Omega\rangle_T \\
&+ \langle\Omega|O_iO_j|\Omega\rangle_T\langle\Omega|O_k|\Omega\rangle_T + \langle\Omega|O_iO_k|\Omega\rangle_T\langle\Omega|O_j|\Omega\rangle_T \\
&+ \langle\Omega|O_kO_j|\Omega\rangle_T\langle\Omega|O_i|\Omega\rangle_T + \langle\Omega|O_i|\Omega\rangle_T\langle\Omega|O_j|\Omega\rangle_T\langle\Omega|O_j|\Omega\rangle_T \\
&\dots
\end{aligned} \tag{D.18}$$

If the O_i's are asymptotic creation or annihilation operators, each truncated vev is proportional to an energy-momentum conservation delta function, since the vacuum state is space-time translation invariant. For a generic choice of particle momenta, only constrained by total energy-momentum conservation, the coefficients of these delta functions are regular, although non-analytic, functions.

However, if the energy-momentum constraint is also satisfied by subsets of the $\boldsymbol{q}$'s and $\boldsymbol{Q}$'s (as is the case for all terms with $K > 1$ in the sum of Eq. (D.17)), singularities in the coefficient functions appear due to the vanishing of some $E_k - H_0 \pm i\epsilon$ denominator in the recursive expansion of Eq. (3.2). In a Lippman-Schwinger approach these singularities correspond to intermediate states whose particles are on their mass-shell. That is, the momenta of all the intermediate particles are fixed by the momentum conservation and the total energy is degenerate with the total energy of the process.

A simple example is provided by the term $K = 2$ for $2 \to 2$ scattering ($n = 2$). This term is proportional to

$$\langle\Omega|A_{\rm out}(\boldsymbol{q}_2)A_{\rm out}(\boldsymbol{q}_2)A_{\rm out}(\boldsymbol{q}_3)A_{\rm out}(\boldsymbol{q}_4)A^\dagger_{\rm in}(\boldsymbol{Q}_1)A^\dagger_{\rm in}(\boldsymbol{Q}_2)A^\dagger_{\rm in}(\boldsymbol{Q}_3)A^\dagger_{\rm in}(\boldsymbol{Q}_4)|\Omega\rangle \tag{D.19}$$

with

$$\boldsymbol{q}_1 + \boldsymbol{q}_2 \simeq \boldsymbol{Q}_1 + \boldsymbol{Q}_2; \qquad E_{\boldsymbol{q}_1} + E_{\boldsymbol{q}_2} \simeq E_{\boldsymbol{Q}_1} + E_{\boldsymbol{Q}_2} \tag{D.20}$$

$$\boldsymbol{q}_3 + \boldsymbol{q}_4 \simeq \boldsymbol{Q}_3 + \boldsymbol{Q}_4; \qquad E_{\boldsymbol{q}_3} + E_{\boldsymbol{q}_4} \simeq E_{\boldsymbol{Q}_3} + E_{\boldsymbol{Q}_4} \tag{D.21}$$

up to corrections of order δ.

It is shown in Appendix B that, if the interaction is given by $\frac{\lambda}{4!\hbar c}\phi^4$, the third order term in the expansion of the amplitude given by Eq. (3.2) contains terms proportional to the product

$$\frac{1}{(E_{\boldsymbol{Q}_1} + E_{\boldsymbol{Q}_2} - E_{\boldsymbol{q}_2} - E_{\boldsymbol{Q}_1+\boldsymbol{Q}_2-\boldsymbol{q}_2} + i\epsilon)(E_{\boldsymbol{q}_3} + E_{\boldsymbol{q}_4} - E_{\boldsymbol{Q}_4} - E_{\boldsymbol{q}_3+\boldsymbol{q}_4-\boldsymbol{Q}_4} + i\epsilon)} \tag{D.22}$$

and analogous ones which apparently diverge if the first two initial particles and the first two final ones fulfill the energy conservation constraint.

It can be checked that, independently of the particular choice of the interaction, a truncated vev of the ordered products of asymptotic creation and annihilation operators $\langle\Omega|\prod_{i=1}^n A_{\rm out}(\boldsymbol{q}_i)\prod_{j=1}^m A^\dagger_{\rm in}(\boldsymbol{Q}_j)|\Omega\rangle_T$ contains a pair of vanishing denominators for each independent subset of initial and final particles fulfilling the energy-momentum conservation constraint. The divergence is actually regulated by the presence of wave packets with a finite width.

In the general case we have

$$\begin{aligned}
\langle\Omega|\prod_{i=1}^{n}(A_{i,\mathrm{out}})^K\prod_{j=1}^{2}(A_{j,\mathrm{in}}^\dagger)^K|\Omega\rangle &= \langle\Omega|\prod_{i=1}^{n}(A_{i,\mathrm{out}})^K\prod_{j=1}^{2}(A_{j,\mathrm{in}}^\dagger)^K|\Omega\rangle_T \\
&+ K\langle\Omega|\prod_{i=1}^{n}(A_{i,\mathrm{out}})^{K-1}\prod_{j=1}^{2}(A_{j,\mathrm{in}}^\dagger)^{K-1}|\Omega\rangle_T\langle\Omega|\prod_{i=1}^{n}A_{i,\mathrm{out}}\prod_{j=1}^{2}A_{j,\mathrm{in}}^\dagger|\Omega\rangle_T + \cdots \\
&+ \Big(\langle\Omega|\prod_{i=1}^{n}A_{i,\mathrm{out}}\prod_{j=1}^{2}A_{j,\mathrm{in}}^\dagger|\Omega\rangle_T\Big)^K
\end{aligned} \tag{D.23}$$

all other possible truncated vev's being zero. By the same procedure that led us to Eq. (D.8), we find

$$\begin{aligned}
\langle\Omega|\prod_{i=1}^{n}(A_{i,\mathrm{out}})^M\prod_{j=1}^{2}(A_{j,\mathrm{in}}^\dagger)^M|\Omega\rangle_T &= \delta^{-4}\left(\frac{\delta}{\sqrt{\pi}}\right)^{\frac{3}{2}M(n+2)} \\
&\prod_{i=1}^{nM}\int d^3\hat{q}_i\, e^{-\frac{\hat{q}_i^2}{2}}\prod_{j=1}^{2M}\int d^3\hat{Q}_j\, e^{-\frac{\hat{Q}_j^2}{2}} \\
&\delta\Big(\sum_{i=1}^{nM}\hat{\boldsymbol{q}}_i+\sum_{i=1}^{2M}\hat{\boldsymbol{Q}}_j\Big)\delta\Big(\sum_{i=1}^{nM}\hat{\boldsymbol{q}}_i\cdot\boldsymbol{v}_i+\sum_{i=1}^{2M}\hat{\boldsymbol{Q}}_j\cdot\boldsymbol{V}_i+O(\delta)\Big) \\
&\sum_{a=0,b}^{M-1}\prod_{l=1}^{2a}\Delta_l^{-1}(a,b,\hat{\boldsymbol{q}},\hat{\boldsymbol{Q}},\boldsymbol{k},\boldsymbol{p})C(a,b,\hat{\boldsymbol{q}},\hat{\boldsymbol{Q}},\boldsymbol{k},\boldsymbol{p}),
\end{aligned} \tag{D.24}$$

where the index a counts the number of vanishing denominators $\Delta_l(a,b,\hat{\boldsymbol{q}},\hat{\boldsymbol{Q}},\boldsymbol{k},\boldsymbol{p})$, b labels the different terms with the same number of vanishing denominators appearing in the expansion and $C(a,b,\hat{\boldsymbol{q}},\hat{\boldsymbol{Q}},\boldsymbol{k},\boldsymbol{p})$ are regular coefficient functions.

As we have seen in the example given above, a generic vanishing denominator has the form

$$\Delta_l(a,b,\boldsymbol{q}',Q',k,p) = \Big(\sum_{i=1}^{n}\boldsymbol{v}_{a,b,i}\cdot\hat{\boldsymbol{q}}_i+\sum_{j=1}^{n}\boldsymbol{V}_{a,b,j}\cdot\hat{\boldsymbol{Q}}_j\Big)\delta+O(\delta^2). \tag{D.25}$$

Hence

$$\langle\Omega|\prod_{i=1}^{n}(A_{i,\mathrm{out}})^M(\prod_{j=1}^{m}A^{\dagger}_{j,\mathrm{in}})^M|\Omega\rangle_T \sim \delta^{-4-2(M-1)}\delta^{\frac{3}{2}M(n+2)} = \delta^{\frac{3}{2}Mn+M-2}. \quad \text{(D.26)}$$

This result implies that in the $\delta \to 0$ limit each term in the sum Eq. (D.23) vanishes as

$$\delta^{\frac{3}{2}Mn+M-2}\delta^{(K-M)(\frac{3}{2}n-1)} = \delta^{\frac{3}{2}Kn-K+2M-2} \quad \text{(D.27)}$$

The sum is therefore dominated by the term $M = 0$:

$$\begin{aligned}\langle\Omega|\prod_{i=1}^{n}(A_{i,\mathrm{out}})^K\prod_{j=1}^{2}(A^{\dagger}_{j,\mathrm{in}})^K|\Omega\rangle &= \left(\langle\Omega|\prod_{i=1}^{n}A_{i,\mathrm{out}}\prod_{j=1}^{2}A^{\dagger}_{j,\mathrm{in}}|\Omega\rangle\right)^K[1+O(\delta^2)]\\ &\sim \delta^{\left(\frac{3}{2}n-1\right)K}[1+O(\delta^2)]\end{aligned} \quad \text{(D.28)}$$

(note that the result Eq. (D.8) is recovered for $K = 1$). As a consequence

$$\langle F|I\rangle - 1 = \langle\Omega|\prod_{i=1}^{n}A_{i,\mathrm{out}}\prod_{j=1}^{2}A^{\dagger}_{j,\mathrm{in}}|\Omega\rangle[1+O(\delta^2)] = \mathcal{A}^{n2}_{kp}[1+O(\delta^{\frac{3}{2}n-1})]. \quad \text{(D.29)}$$

This completes the proof.

The representation Eq. (D.29) of the generic transition amplitude will prove particularly suited to identify the asymptotic properties of the field operator, which is a crucial step in the construction of the semiclassical approximation to the scattering amplitude.

Appendix E
Scattering from an External Density

A simple and interesting application of our formulae Eqs. (3.124) and (3.129) is the computation of scattering amplitudes in a model with interaction given by

$$\mathcal{L}_I = \frac{g(x)}{2}\phi^2(x). \tag{E.1}$$

In this model, the field equations are linear, and the function $g(x)$ plays a role which is similar to that of a potential in the Schrödinger equation. The solution of Eq. (3.129) is

$$\begin{aligned} S\left[\phi^{(\mathrm{as})}\right] &= -\frac{1}{2}\sum_{n=0}^{\infty}(-1)^n \int d^4x\, g(x)\phi^{(\mathrm{as})}(x)\left(\left(\Delta * g\right)^n \phi^{(\mathrm{as})}\right)(x) \\ &= -\frac{1}{2}\sum_{n=0}^{\infty}(-1)^n \int d^4x \prod_{i=1}^{n} dy_i \phi^{(\mathrm{as})}(x) g(x)\Delta(x-y_1)g(y_1) \\ &\quad \Delta(y_1-y_2)g(y_2)\cdots\Delta(y_{n-1}-y_n)g(y_n)\phi^{(\mathrm{as})}(y_n). \end{aligned} \tag{E.2}$$

Thus, S is quadratic in $\phi^{(\mathrm{as})}(x)$, and describes three possible processes: two-particle annihilation, pair production, and single-particle scattering. To first order in g, one has

$$S\left[\phi^{(\mathrm{as})}\right] = -\frac{1}{2}\int d^4x\, g(x)\left(\phi^{(\mathrm{as})}(x)\right)^2, \tag{E.3}$$

and therefore the scattering amplitude

$$S_{p\to k}[g] = -\int d^4x\, g(x)\frac{e^{i(k-p)\cdot x}}{2(2\pi)^3\sqrt{E_p E_k}} \equiv -\pi\frac{\tilde{g}(p-k)}{\sqrt{E_p E_k}} \tag{E.4}$$

while the annihilation amplitude for a pair of particles with momenta p_1 and p_2 is

C. M. Becchi and G. Ridolfi, *An Introduction to Relativistic Processes and the Standard Model of Electroweak Interactions*, UNITEXT for Physics,
DOI: 10.1007/978-3-319-06130-6,

$$S_{\boldsymbol{p}_1,\boldsymbol{p}_2\to 0}[g] = -\pi \frac{\tilde{g}(p_1+p_2)}{\sqrt{E(p_1)E(p_2)}}. \tag{E.5}$$

Finally, the pair production amplitude is

$$S_{0\to \boldsymbol{k}_1,\boldsymbol{k}_2}[g] = -\pi \frac{\tilde{g}(-k_1-k_2)}{\sqrt{E_{k_1}E_{k_2}}}. \tag{E.6}$$

We now consider the case

$$g(x) = g_1(x) + g_2(x), \tag{E.7}$$

with

$$\theta(x_2^0 - x_1^0) g_1(x_1) g_2(x_2) = g_1(x_1) g_2(x_2), \tag{E.8}$$

that is, g_1 acts before g_2. Selecting in the scattering amplitude due to $g(x)$ the first order terms in g_1 and g_2, one has

$$\begin{aligned} S_{1,2} &= \int d^4x\, g_1(x)\phi^{(\mathrm{as})}(x)\left(\Delta \circ g_2\phi^{(\mathrm{as})}\right)(x) \\ &= \int d^4x_1 d^4x_2\, g_1(x_1)\phi^{(\mathrm{as})}(x_1)\Delta(x_1-x_2)g_2(x_2)\phi^{(\mathrm{as})}(x_2). \end{aligned} \tag{E.9}$$

Replacing

$$\phi^{(\mathrm{as})}(x) = \frac{e^{ik\cdot x}}{\sqrt{2E_{\boldsymbol{k}}(2\pi)^3}} + \frac{e^{-ip\cdot x}}{\sqrt{2E_{\boldsymbol{p}}(2\pi)^3}}, \tag{E.10}$$

and recalling Eqs. (E.8) and (3.95), we find

$$\begin{aligned} S_{1,2} &= \frac{\pi^2}{\sqrt{E_{\boldsymbol{p}}E_{\boldsymbol{k}}}} \int \frac{d\boldsymbol{q}}{E_{\boldsymbol{q}}} \left[\tilde{g}_1(\boldsymbol{p}-\boldsymbol{q})\tilde{g}_2(\boldsymbol{q}-\boldsymbol{k}) + \tilde{g}_1(-\boldsymbol{k}-\boldsymbol{q})\tilde{g}_2(\boldsymbol{q}+\boldsymbol{p})\right] \\ &= \int d\boldsymbol{q} \left[S_{\boldsymbol{p}\to\boldsymbol{q}}[g_1]S_{\boldsymbol{q}\to\boldsymbol{k}}[g_2] + S_{0\to\boldsymbol{q},\boldsymbol{k}}[g_1]S_{\boldsymbol{p},\boldsymbol{q}\to 0}[g_2]\right]. \end{aligned} \tag{E.11}$$

This shows that the scattering amplitude appears as the sum of two terms. The first describes the scattering due to g_1 from $\boldsymbol{p}$ to $\boldsymbol{q}$, followed by the scattering due to g_2 from $\boldsymbol{q}$ to $\boldsymbol{k}$. The second term describes creation from the vacuum of a pair with momenta $\boldsymbol{k}$ and $\boldsymbol{q}$, followed by the annihilation of the initial particle with momentum $\boldsymbol{p}$ with that with momentum $\boldsymbol{q}$. Thus, the scattering process factorizes into the contributions due to g_1 and g_2 consistently with the causal order.

Appendix F
Dirac Matrices

Most calculations involving Dirac matrices can be performed using only their general properties,

$$\{\gamma_\mu, \gamma_\nu\} = 2I g_{\mu\nu}; \qquad \gamma_5 = -i\gamma_0\gamma_1\gamma_2\gamma_3 = -\frac{i}{4!}\,\epsilon^{\mu\nu\rho\sigma}\,\gamma_\mu\gamma_\nu\gamma_\rho\gamma_\sigma,$$
$$\gamma_\mu^\dagger = \gamma_0\,\gamma_\mu\,\gamma_0, \tag{F.1}$$

with no reference to a specific representation. Some immediate consequences of Eq. (F.1) are

$$\mathrm{Tr}\,\gamma_\mu\gamma_\nu = 4\,g_{\mu\nu}; \qquad \{\gamma_\mu, \gamma_5\} = 0; \qquad \gamma_5^2 = I, \tag{F.2}$$

and the identities

$$\gamma^\mu\,\gamma^\alpha\,\gamma_\mu = -2\gamma^\alpha \tag{F.3}$$
$$\gamma^\mu\,\gamma^\alpha\,\gamma^\beta\,\gamma_\mu = 4\,g^{\alpha\beta} \tag{F.4}$$
$$\gamma^\mu\,\gamma^\alpha\,\gamma^\beta\,\gamma^\gamma\,\gamma_\mu = -2\gamma^\gamma\,\gamma^\beta\,\gamma^\alpha. \tag{F.5}$$

The trace of the product of an odd number of γ vanishes. Indeed,

$$\gamma_\mu = -\gamma_5\,\gamma_\mu\,\gamma_5, \tag{F.6}$$

and therefore

$$\begin{aligned}\mathrm{Tr}\,\gamma_{\mu_1}\cdots\gamma_{\mu_{2n+1}} &= (-1)^{2n+1}\mathrm{Tr}\,\left(\gamma_5\gamma_{\mu_1}\gamma_5\right)\cdots\left(\gamma_5\gamma_{\mu_{2n+1}}\gamma_5\right)\\ &= -\mathrm{Tr}\,\gamma_{\mu_1}\cdots\gamma_{\mu_{2n+1}},\end{aligned} \tag{F.7}$$

where we have used the circular property of the trace. It is easy to prove that

C. M. Becchi and G. Ridolfi, *An Introduction to Relativistic Processes and the Standard Model of Electroweak Interactions*, UNITEXT for Physics,
DOI: 10.1007/978-3-319-06130-6, © Springer International Publishing Switzerland 2014

$$\mathrm{Tr}\,\gamma^\mu\gamma^\nu\gamma^\rho\gamma^\sigma = 4(g^{\mu\nu}g^{\rho\sigma} - g^{\mu\rho}g^{\nu\sigma} + g^{\mu\sigma}g^{\nu\rho}) \tag{F.8}$$

$$\mathrm{Tr}\,\gamma^\mu\gamma^\nu\gamma^\rho\gamma^\sigma\gamma_5 = 4i\epsilon^{\mu\nu\rho\sigma}. \tag{F.9}$$

In Sect. 6.1 we have introduced a particular representation of the Dirac matrices:

$$\gamma_\mu = \begin{pmatrix} 0 & \sigma_\mu \\ \bar{\sigma}_\mu & 0 \end{pmatrix}. \tag{F.10}$$

Different representations of the γ matrices are related by similarity transformations on spinor fields. A representation which is often used (especially in applications that involve the non-relativistic limit) is the so-called standard representation:

$$\gamma^0 = \begin{pmatrix} I & 0 \\ 0 & -I \end{pmatrix}, \quad \gamma^i = \begin{pmatrix} 0 & -\sigma^i \\ \sigma^i & 0 \end{pmatrix}, \quad \gamma^5 = \begin{pmatrix} 0 & I \\ I & 0 \end{pmatrix}. \tag{F.11}$$

Appendix G
Violation of Unitarity in the Fermi Theory

In this Appendix, we show that unitarity of the $\mathcal{S}$ matrix is violated in the Fermi theory of weak interactions. We rewrite the unitarity constraint, Eq. (4.49), for $i = j$:

$$\sum_f \int d\phi_{n_f}(P_i; k_1^f, \ldots, k_{n_f}^f) \, |\mathcal{M}_{if}|^2 = -2 \, \mathrm{Im} \, \mathcal{M}_{ii}, \tag{G.1}$$

which is the so-called optical theorem: the total cross section for the process $i \to f$, summed over all possible final states f, is proportional to the imaginary part of the forward invariant amplitude $\mathcal{M}_{ii}$.

Let us now assume that i is a state of two massless particles with momenta p_1, p_2; furthermore, let us assume that only $2 \to 2$ processes are allowed. Under these conditions, the states f are also two-particle states, and the amplitudes $\mathcal{M}_{if}$ depend on the initial and final states through the two independent Mandelstam variables s, t:

$$\mathcal{M}_{if} = \mathcal{M}(s, t), \tag{G.2}$$

where

$$s = (p_1 + p_2)^2, \qquad t = (p_1 - k_1)^2. \tag{G.3}$$

In the center-of-mass frame,

$$t = -\frac{s}{2}(1 - \cos\theta) \;\to\; \cos\theta = 1 + \frac{2t}{s}, \tag{G.4}$$

where θ is the scattering angle. Thus, for a given value of the center-of-mass squared energy s, the amplitude $\mathcal{M}(s, t)$ is a function of $\cos\theta$ only, and can be expanded on the basis of the Legendre polynomials

$$P_J(z) = \frac{1}{J!2^J} \frac{d^J}{dz^J} (z^2 - 1)^J. \tag{G.5}$$

C. M. Becchi and G. Ridolfi, *An Introduction to Relativistic Processes and the Standard Model of Electroweak Interactions*, UNITEXT for Physics,
DOI: 10.1007/978-3-319-06130-6,

The Legendre polynomials obey the orthogonality conditions

$$\int_{-1}^{1} dz\, P_J(z)\, P_K(z) = \frac{2}{2J+1}\,\delta_{JK} \tag{G.6}$$

and the normalization conditions

$$P_J(1) = 1. \tag{G.7}$$

We find

$$\mathcal{M}(s,t) = 16\pi \sum_J (2J+1)\, a_J(s)\, P_J(\cos\theta), \tag{G.8}$$

where the partial-wave amplitudes $a_J(s)$ are given by

$$a_J(s) = \frac{1}{32\pi} \int_{-1}^{1} d\cos\theta\, P_J(\cos\theta)\, \mathcal{M}(s,t). \tag{G.9}$$

Replacing Eq. (G.8) in the l.h.s. of Eq. (G.1) we get

$$\begin{aligned}
&\int \frac{d^3k_1}{(2\pi)^3\, 2E_{k_1}} \frac{d^3k_2}{(2\pi)^3\, 2E_{k_2}} (2\pi)^4 \delta^{(4)}(p_1 + p_2 - k_1 - k_2)\, |\mathcal{M}(s,t)|^2 \\
&= \frac{1}{16\pi} \int_{-1}^{1} d\cos\theta \\
&\left[16\pi \sum_J (2J+1)\, a_J(s)\, P_J(\cos\theta)\right]\left[16\pi \sum_K (2K+1)\, a_K^*(s)\, P_K(\cos\theta)\right] \\
&= 32\pi \sum_J (2J+1)\, |a_J(s)|^2,
\end{aligned} \tag{G.10}$$

while the r.h.s. is given by

$$-2\,\mathrm{Im}\,\mathcal{M}(s,0) = -32\pi \sum_J (2J+1)\, \mathrm{Im}\, a_J(s), \tag{G.11}$$

where we have set $t = 0$, or equivalently $\cos\theta = 1$, as appropriate for a forward amplitude, and we have used the normalization condition (G.7). Therefore, the unitarity constraint Eq. (G.1) requires

$$|a_J(s)|^2 = -\mathrm{Im}\, a_J(s) \tag{G.12}$$

for all partial amplitudes. Equation (G.12) provides the unitarity bound

$$|a_J(s)| \leq 1. \tag{G.13}$$

Let us now consider a specific process, namely

$$e^-(p_1) + \nu_\mu(p_2) \to \mu^-(k_1) + \nu_e(k_2) \tag{G.14}$$

within the Fermi theory. The relevant amplitude is

$$\mathcal{M}(s,t) = -\frac{G_F}{\sqrt{2}}\,\bar{u}(k_2)\,\gamma^\alpha(1-\gamma_5)\,u(p_1)\,\bar{u}(k_1)\,\gamma_\alpha(1-\gamma_5)\,u(p_2), \tag{G.15}$$

where all lepton masses have been neglected. This gives

$$\begin{aligned}\sum_{\text{pol}} |\mathcal{M}(s,t)|^2 &= 2G_F^2\,\mathrm{Tr}\left[\gamma^\alpha\,\not{p}_1\,\gamma^\beta(1-\gamma_5)\,\not{k}_2\right]\,\mathrm{Tr}\left[\gamma_\alpha\,\not{p}_2\,\gamma_\beta(1-\gamma_5)\,\not{k}_1\right]\\ &= 32G_F^2 s^2.\end{aligned} \tag{G.16}$$

We see that only the partial amplitude $a_0(s)$ is nonzero, since there is no t dependence at all. Using the definition Eq. (G.9) we obtain

$$|a_0(s)| = \frac{G_F\,s}{2\sqrt{2}\pi}. \tag{G.17}$$

The unitarity bound Eq. (G.13) is therefore violated for

$$\sqrt{s} \geq \sqrt{\frac{2\sqrt{2}\pi}{G_F}} \simeq 875\,\text{GeV}. \tag{G.18}$$

The total cross section obtained from Eq. (G.16),

$$\sigma = \frac{G_F^2 s}{2\pi}, \tag{G.19}$$

grows linearly with the squared center-of-mass energy s.

In the standard model, the same amplitude involves the exchange of a virtual W boson with mass m_W and coupling $g/(2\sqrt{2})$ to left-handed fermions. The standard model squared amplitude is obtained from the result in Eq. (G.16) by the replacement

$$-\frac{G_F}{\sqrt{2}} \to \frac{g^2}{8}\frac{1}{t-m_W^2} = \frac{G_F}{\sqrt{2}}\frac{m_W^2}{t-m_W^2}. \tag{G.20}$$

We get

$$\sum_{\text{pol}} \left| \mathcal{M}^{\text{SM}}(s,t) \right|^2 = 64 G_F^2 s^2 \left(\frac{m_W^2}{t - m_W^2} \right)^2 . \tag{G.21}$$

The total cross section is now given by

$$\sigma^{\text{SM}} = \frac{G_F^2 s}{2\pi} \frac{m_W^2}{s + m_W^2}, \tag{G.22}$$

that reduces to the result obtained in the Fermi theory, Eq. (G.19), for $s \ll m_W^2$. In this case, however, the linear growth of the cross section with s is cut off at $s \sim m_W^2$. At very large energy,

$$\sigma^{\text{SM}} \to \frac{G_F^2 m_W^2}{2\pi}. \tag{G.23}$$

The value of m_W is related to the size of the coupling g through $G_F/\sqrt{2} = g^2/(8m_W^2)$. If m_W were close to the energy at which the Fermi theory breaks down, about 900 GeV, then g would take a value close to 10, far from the perturbative domain. The fact that the measured value m_W is instead much smaller, $m_W \simeq 80$ GeV, is a signal of the fact that a theory of weak interactions with an intermediate vector boson can be treated perturbatively: indeed, in this case we get $g \sim 0.7$.

Index

C. M. Becchi and G. Ridolfi, *An Introduction to Relativistic Processes and the Standard Model of Electroweak Interactions*, UNITEXT for Physics,
DOI: 10.1007/978-3-319-06130-6, © Springer International Publishing Switzerland 2014

Printed in the United States
By Bookmasters